Enhancing Dynamic Command and Control of Air Operations Against Time Critical Targets

Myron Hura
Gary McLeod
Richard Mesic
Philip Sauer
Jody Jacobs
Daniel Norton
Thomas Hamilton

Prepared for the United States Air Force

RAND
Project AIR FORCE

The research reported here was sponsored by the United States Air Force under Contract F49642-01-C-0003. Further information may be obtained from the Strategic Planning Division, Directorate of Plans, Hq USAF.

ISBN: 0-8330-3131-7

RAND is a nonprofit institution that helps improve policy and decisionmaking through research and analysis. RAND® is a registered trademark. RAND's publications do not necessarily reflect the opinions or policies of its research sponsors.

Published 2002 by RAND
1700 Main Street, P.O. Box 2138, Santa Monica, CA 90407-2138
1200 South Hayes Street, Arlington, VA 22202-5050
201 North Craig Street, Suite 102, Pittsburgh, PA 15213
RAND URL: http://www.rand.org/
To order RAND documents or to obtain additional information, contact Distribution Services: Telephone: (310) 451-7002;
Fax: (310) 451-6915; Email: order@rand.org

PREFACE

During the 1990s, air operations in Southwest Asia and the Balkans against mobile air defenses, tactical ballistic missiles and their launch means, and fielded forces operating in small units and intermingled with civilian populations served to highlight shortfalls in U.S. dynamic command and control and battle management (DC2BM). The objective of this study is to help the Air Force improve DC2BM of intelligence, surveillance, and reconnaissance assets and weapon systems in air operations against adversaries that are increasingly employing asymmetrical strategies and denial and deception techniques and leveraging advances in technologies to reduce exposure of many of their critical military systems to U.S. air assets.

The focus of this study is on the ability of the Air Force to effectively attack time critical targets (TCTs) and the extent to which its successes and failures in this area can be attributed to dynamic command and control and battle management capabilities (or their lack). Specifically, the study identifies key DC2BM functions and capabilities necessary to conduct operations against TCTs; surveys new DC2BM programs and initiatives to determine whether they provide the desired capabilities; and develops suggested actions to address identified shortfalls.

This report is an abridged version of a longer, limited-distribution report (Hura, McLeod, et al., 2002) that documents the analysis in much greater detail. The present report focuses on the DC2BM shortfalls identified in the detailed analysis and presents suggested areas for improving DC2BM.

This research was sponsored by the Air Force Director of Intelligence, Surveillance, and Reconnaissance (USAF/XOI) and the Air Force Director for Command and Control (USAF/XOC), and was performed in the Aerospace Force Development program of Project AIR FORCE. The principal research was completed in September 2000.

The research should be of interest to Air Force and Department of Defense personnel involved in developing and fielding enhanced command and control and battle management (C2BM) systems, and to operators developing improved tactics, techniques, and procedures for force-level and unit-level air operations.

Project AIR FORCE

Project AIR FORCE, a division of RAND, is the Air Force federally funded research and development center for studies and analysis. It provides the Air Force with independent analyses of policy alternatives affecting the development, employment, combat readiness, and support of current and future aerospace forces. Research is performed in four programs: Aerospace Force Development; Manpower, Personnel, and Training; Resource Management; and Strategy and Doctrine.

CONTENTS

TABLE

SUMMARY

During the 1990s, air operations against adversary mobile force elements (surface-to-air missile defenses, tactical ballistic missile launchers, and fielded forces operating in small units) that employed camouflage, concealment, and deception techniques and shoot-and-scoot tactics produced disappointing results. The objective of this study is to help the Air Force develop enhanced dynamic command and control and battle management (DC2BM) of intelligence, surveillance, and reconnaissance (ISR) assets and shooter assets in air operations against these time critical targets (TCTs).

ANALYTIC APPROACH

Our analytical approach considered DC2BM in an overall C2BM context. Both *deliberate* and *dynamic* C2BM practices are employed to monitor, assess, plan, and execute air and space operations. For example, the traditional air tasking order cycle is predominantly deliberate (that is, targets are selected and weapons and targets are paired up to 72 hours in advance of the missions), but certain preplanned forces are designated to conduct dynamic operations such as defensive and offensive counterair and close air support. DC2BM is therefore not new. What is new is the increased emphasis on less-traditional TCTs and enemy tactics that require the application of increasingly dynamic C2BM.

We conducted discussions with participants and reviewed documents from recent operations (Desert Storm, Allied Force, Southern Watch, and Northern Watch) to gain insights into key factors that shape DC2BM and to identify key shortfalls that surfaced during these operations.

Next, we reviewed ongoing and planned Air Force and Department of Defense (DoD) programs and initiatives and new technologies and practices, to identify potential solution alternatives. The combination of shortfalls identified in recent operations and lessons learned by potential adversaries and their responses shaped our examination of options to address those shortfalls. We then assessed the relative military value of these alternatives in mission-level analyses of representative defensive and offensive counterair (DCA/OCA), theater missile defense (TMD), suppression of enemy air defenses (SEAD), and interdiction operations.

In our analyses of shortfalls and options to address those shortfalls, we decomposed DC2BM into six key functional areas:

- Integrated tasking and rapid retasking of sensors and related processing, exploitation, and dissemination.

- Timely integration (correlation and fusion) of information derived from multiple sources.

- Rapid target development and target nomination.

- Rapid weapon and target pairing.

- Timely decision and attack order dissemination.

- Rapid assessment of effects of weapon delivery.

We used these areas as a framework to help us identify individual mission shortfalls and integrate the shortfalls across the four mission areas described in the next section.

Finally, taking into account planned ISR investments and weapon systems investments, we identified top-level issues that the Air Force and DoD should consider in improving the DC2BM of future air operations.

MISSION-AREA DC2BM SHORTFALLS AND SUGGESTED ACTIONS

Our review and our qualitative and quantitative analyses of past air operations and future projections identified four mission areas in which existing DC2BM capabilities are inadequate:

- Counterair operations against cruise missiles.

- TMD counterforce operations against tactical ballistic missile (TBM) transporter-erector-launchers (TELs).

- SEAD in the context of strike missions against targets defended by advanced air defenses.

- Interdiction of small-unit ground forces intermingled with the civilian population.

Counterair Operations Against Cruise Missiles

Nowhere is DC2BM more critical than in the area of air-to-air engagements, where the speed of operations dictates close coordination among the various air defense elements. Except for concerns regarding combat identification, the Air Force is well-equipped and well-trained to conduct counterair operations against a conventional military aircraft threat (e.g., fighters, bombers, ISR aircraft) posed by likely future adversaries in major theater wars and lesser-intensity conflicts.[1]

However, both enhanced DC2BM and sensor improvements will be needed against adversaries that could employ stealthy cruise missiles. In particular, initiatives that support development of good situational awareness and an accurate air picture are needed to support effective and efficient air-to-air engagements[2] in a crowded airspace that contains coalition military aircraft (both conventional and stealthy) and enemy cruise missiles. Integration of the Radar System

---

[1]Asymmetrical threats such as hijacked commercial airliners by terrorists in suicide missions pose added DC2BM challenges that should be examined in future research.

[2]We did not examine counterforce operations against cruise missile launchers and support facilities because most of the DC2BM challenges in such operations are comparable to those in the TMD counterforce mission.

Improvement Program (RSIP) on the Airborne Warning and Control System (AWACS) will improve cruise missile detection.[3] A new tracker, improved interrogation capabilities, integration and fusion of onboard and offboard sensor data, and new computers and displays are needed to improve tracking and identification and to reduce C2BM process timelines.

Theater Missile Defense Counterforce Operations

Allied experience with Iraqi Scuds during Desert Storm led to increasing emphasis on systems and operational concepts for dealing with this category of TCTs. Our analysis shows that current DC2BM capabilities are still inadequate to support prelaunch and postlaunch counterforce operations against TBMs and their TELs. Improvements in all six key functional areas are essential to ensure that the DC2BM process is completed in a tight, threat-driven timeline.[4] Shortening current timelines will require refinement of the current concept of operations (CONOPS) and creation of a C2BM organizational structure, with refined tactics, techniques, and procedures (TTP), expert personnel, and extensive automation (tools and applications).

Future ISR improvements will also be required to significantly improve both prelaunch and postlaunch counterforce effectiveness. For postlaunch operations, these ISR improvements include high-resolution, focused-look (because they are cued) imaging sensors that are all-weather and day/night capable, which suggests radar sensors with synthetic aperture radar (SAR) modes. In addition, they must be deep-look, long-dwell, and survivable, which places significant constraints on both the sensor and the sensor platform. To track the TBM TEL and related components once they begin to move, possibly in urban clutter, requires robust tracking algorithms

---

[3]RSIP is the only program specifically mentioned in the summary because the Air Force is currently buying the upgrade and is installing it on AWACS. Other efforts to improve capabilities against TCTs are merely initiatives and experiments that require more development; they are not funded acquisition programs.

[4]A timeline approaching 10 minutes or less—measured from the time of initial detection of a possible TCT to order issuance to a strike aircraft—is required. This timeline does not include the flight times of retasked sensors and strike aircraft to possible targets.

and a radar with a ground moving target indication mode, preferably with high range resolution capability for target classification. For prelaunch operations, a focused-look sensor is insufficient; such operations require wide-area, high-resolution SAR sensors with rapid scan rates to search the deployment area and enhanced automatic target recognition algorithms to limit the number of false alarms.

Without these advanced ISR capabilities, TMD counterforce operations will continue to be relatively ineffective. Conversely, personnel responsible for DC2BM improvement programs must ensure that their efforts are supportive of future ISR improvements.

Suppression of Enemy Air Defenses

The threat presented by older air defense systems has not unduly constrained strike and interdiction missions. In air operations during the past 10 years, U.S. and coalition air assets have accomplished their missions with minimal attrition (less than one-tenth of one percent per sortie in Desert Storm and less than one-hundredth of one percent per sortie in Allied Force). Threat avoidance, judicious tactics, self-defense capabilities, suppression (jamming, anti-radiation missiles, air-launched decoys), and the associated DC2BM have been effective.

However, DC2BM capabilities must be improved for SEAD operations against advanced surface-to-air missile systems such as SA-10s and SA-20s, to counter their increased lethality, robustness, and mobility. Otherwise, strike and interdiction missions in future air campaigns may be severely constrained. As with the TMD counterforce mission, improvements in all six DC2BM functional areas are needed to shorten response timelines. In addition to automated tools, new C2BM organization, refined CONOPS and TTP, and expert personnel are essential enablers to perform the DC2BM functions. These processes must also be adapted to incorporate new information warfare techniques.

In addition, ISR shortfalls must be addressed. Many of the ISR solutions to the TMD problems will be useful in the SEAD mission, but specialized cueing capabilities (e.g., precision signals intelligence) will also be required. One ISR shortfall will be difficult to overcome—the inability to detect, identify, track, and target anti-aircraft artillery and man-portable surface-to-air missiles—and this will continue to constrain low-level air operations.

Interdiction of Small-Unit Ground Forces

Our analysis of interdiction operations against ground forces operating in large formations (for example, during the halt phase of an air campaign to stop an invading army) suggests that existing DC2BM capabilities are effective in this area.

However, existing DC2BM and ISR capabilities are inadequate for air operations against ground forces operating in small units intermingled with the civilian population. In such operations, it is often necessary to have positive target identification before attack and high-quality information to determine the effect of attacks. Neither airborne forward air controllers, nor offboard airborne sensors, nor decisionmakers in the air operations center are currently capable of positive identification of small-unit ground forces—especially dismounted units—intermingled with civilian populations. Nor can they confirm that air-delivered weapons were aimed at and struck the right targets.

A high-resolution, focused-look sensor that is capable of distinguishing small-unit ground forces from friendly civilians is an important missing element. As in the other mission areas, the sensors must be all-weather, day/night capable, deep-look, long-dwell, and survivable. All these objectives may not be achievable solely by using air and/or space sensor assets. Army and special operations forces possess capabilities to support such ISR operations, but their employment in certain hostile environments may not be worth the risk.

Our research suggests that developing CONOPS, TTP, ISR capabilities, and weapon systems with characteristics comparable to police force operations offers a reasonable approach for addressing this very difficult mission, at least in low-threat environments or in hostile environments in which the benefit outweighs the risk. However, differences between the operational situations faced by the police and the military will, in all likelihood, affect the military's specific implementation and rules of engagement.

TOP-LEVEL SUGGESTED ACTIONS TO ENHANCE DC2BM

We suggest that DC2BM improvements focus on the following:

- Refining CONOPS and TTP and developing an end-to-end, scalable functionality for operations against TCTs that integrates ISR, C2BM, and weapons systems.

- Building a robust, collaborative, distributed environment (tools for collaboration; on-demand, sufficient, and assured communications; a flexible network and server architecture with responsive operating protocols; an effective network manager; and an empowered information manager).

- Extensively automating the applications for performing the six DC2BM functions discussed above.

- Synchronizing the new applications with a Web-enabled and wrapper-enabled Theater Battle Management Core System.

Without such improvements, the C2BM response will be slow. The DC2BM functions must be completed in times approaching 10 minutes or less for successful operations against many classes of TCTs.

In developing a flexible TCT functional capability and selecting specific tools to support the timely performance of the six DC2BM functional areas, we identified four important factors that should be considered. We highlight two here:[5]

- Tensions between planned and dynamically tasked missions.

- Balanced and synchronized investments among DC2BM, ISR, aircraft, and weapons.

New DC2BM improvements should be designed so that they do not jeopardize Air Force capabilities to execute deliberately planned strike and air superiority missions that are necessary to achieve the operational objectives of the air campaign plan. Unwise dynamic reallocation of weapons systems from preplanned missions (missions that nominally have a high likelihood of success) to TCTs

---

[5]The other two factors are (1) jointness and defense in depth, and (2) robustness and flexibility.

(missions that currently have a much lower likelihood of success) may disrupt the intended battle rhythm and the timely accomplishment of intended campaign objectives. Thus, new DC2BM programs must interact seamlessly with key components (i.e., those that are integral to deliberate operations) of the Theater Battle Management Core System.

In addition, investments in DC2BM must be balanced and synchronized with those of ISR, aircraft, and weapons to ensure that desired counter-TCT capabilities are met. Without improved ISR and more-responsive weapon systems, enhanced DC2BM capabilities will not substantially improve air power performance against TCTs. Conversely, without enhanced DC2BM capabilities, the full potential of new ISR, aircraft, and weapon system investments will not be realized. This clearly suggests that high-level attention should be focused on integrating ISR, C2BM, and weapons—from doctrine; through C2BM organization, CONOPS, TTP, and systems; to trained personnel.

ACKNOWLEDGMENTS

We gratefully acknowledge the steadfast support we received from our study points of contact, Lieutenant Colonel Warren Abraham (AF/XOII), Lieutenant Colonel David Nichols (AF/XOCI), and Major Kevin Fox (AF/XOCI) at the Air Staff. We also thank Colonel Joseph May (AC2ISRC/A-3) and members of his staff for providing useful and relevant information. Lieutenant Colonel David Jones (AC2ISRC/C2N) provided useful insights on approaches and planned programs for addressing time critical targets.

Several RAND colleagues made valuable contributions to this research. Lewis Jamison helped us understand sensors on board fighters. Daniel Gonzales, David Matonick, and Louis Moore helped us better understand interdiction operations against fielded ground forces. Barry Wilson, Jeff Hagen, and Bart Bennett helped us with our SEAD analysis. Lieutenant Colonel Vickie Woodard, while an Air Force Executive Fellow at RAND, provided useful insights into recent military operations, in particular, Operation Southern Watch. Finally, we thank Leland Joe and Jeff Hagen for their thoughtful reviews of the draft report. Of course, we alone are responsible for any errors of omission or commission.

ACRONYMS

AADC	area air defense commander
AB	Air Base
ABCCC	Airborne Battlefield Command and Control Center
AC2ISRC	Aerospace Command and Control and Intelligence, Surveillance, and Reconnaissance Center
ACC	Air Combat Command
ACE	airborne command element
ACW	air control wing
AFB	Air Force Base
AOACMT	Attack Operations Against Critical Mobile Targets
AOC	air operations center
ATO	air tasking order
AWACS	Airborne Warning and Control System
BM	battle management
BMDO	Ballistic Missile Defense Organization
C2	command and control
C2BM	command and control and battle management
C2ISR	command, control, intelligence, surveillance, and reconnaissance

C3I	command, control, communications, intelligence
C3ISR	command, control, communications, intelligence, surveillance, and reconnaissance
C4ISR	command, control, communications, computers, intelligence, surveillance, and reconnaissance
CAOC	combined air operations center
CEC	Cooperative Engagement Capability
CONOPS	concept of operations
CONUS	continental United States
cross-domain	across sensor platform domains (e.g., space, air, sea, land)
DASD (C3ISR&S)	Deputy Assistant Secretary of Defense (C3ISR and Space)
DCA	defensive counterair
DC2BM	dynamic C2BM
DoD	Department of Defense
EFX	expeditionary force experiment
FSRD	family of systems requirements document
GMTI	ground moving target indication
GPS	Global Positioning System
HF	high frequency
HRR	high range resolution
HTML	hypertext markup language
IFF	identification friend or foe
INT	intelligence [discipline]
IOC	initial operational capability
IPB	intelligence preparation of the battlespace

ISR	intelligence, surveillance, and reconnaissance
IW	information warfare
JCTN	joint composite tracking network
JDN	joint data network
JEFX	joint expeditionary force experiment
JFACC	joint force air component commander
JIGI	JSTARS Imagery Geolocation Improvement
JIPTL	joint integrated prioritized target list
JPO	joint project office
JSTARS	Joint Surveillance [and] Target Attack Radar System
JT&E	joint test and evaluation
JTAMDO	Joint Theater Air and Missile Defense Organization
JTF	joint task force
JTIDS	Joint Tactical Information Distribution System
JWID	Joint Warrior Interoperability Demonstration
MC2A	multisensor command and control aircraft
MMA	multimission aircraft
MP-RTIP	Multi-Platform Radar Technology Insertion Program
multi-INT	multiple intelligence [disciplines]
OASD(C3I)	Office of the Assistant Secretary of Defense (C3I)
OCA	offensive counterair
OSD	Office of the Secretary of Defense
PED	processing, exploitation, and dissemination
RSIP	Radar System Improvement Program
RTIP	Radar Technology Insertion Program
SA	situational awareness

SAM	surface-to-air missile
SAR	synthetic aperture radar
SEAD	suppression of enemy air defenses
single-INT	single intelligence [discipline]
TADIL	tactical digital information link
TAMD	theater air and missile defense
TBM	theater ballistic missile
TBMCS	Theater Battle Management Core System
TCT	time critical target
TEL	transporter-erector-launcher
TMD	theater missile defense
TST	time sensitive target
TTP	tactics, techniques, and procedures
UAV	unmanned aerial vehicle
UHF	ultra high frequency
U.S.	United States
USAF	U.S. Air Force
USAFE	U.S. Air Forces in Europe
USJFCOM	U.S. Joint Forces Command
VHF	very high frequency
WMD	weapons of mass destruction
XML	extensible markup language

Chapter One

INTRODUCTION

In air operations in Southwest Asia and the Balkans during the 1990s, the United States proved the effectiveness of its strike and interdiction capabilities against fixed targets and stationary force elements. Recognizing the vulnerability of exposed stationary targets, adversaries have responded by (1) taking advantage of advances in new weapon and information technologies to develop more-mobile force elements; (2) enhancing camouflage, concealment, and deception techniques to minimize the exposure of their high-value assets; and (3) employing asymmetrical strategies. Such responses frustrated U.S. counterforce operations against Iraqi tactical ballistic missiles in Operation Desert Storm. In Operation Allied Force, the Serbian military used similar means to limit the effectiveness of U.S. strike and interdiction capabilities[1] against Serbian fielded forces in Kosovo, which were operating from hiding places and intermingling with the civilian population. Attacks to suppress enemy air defenses through strikes on mobile air defense systems, which employed emission control and used shoot-and-scoot tactics, also produced somewhat disappointing results.

STUDY OBJECTIVE

The U.S. military recognizes the difficult challenge of prosecuting these sorts of targets, generally referred to as time sensitive targets

---

[1] We use a broad definition of capabilities to include intelligence, surveillance, and reconnaissance (ISR); command and control (C2); battle management (BM); and communications; as well as weapon systems.

1

(TSTs) or time critical targets (TCTs). Numerous initiatives are under way to improve attack effectiveness; the challenge is to develop and integrate those that offer the best payoff and to allocate sufficient resources to field them. The objective of this study is to help the Air Force enhance its dynamic command and control and battle management (DC2BM) of air assets in operations against TSTs and TCTs in major theater wars and lesser intensity conflicts.

TIME SENSITIVE AND TIME CRITICAL TARGETS

We used the following definitions of time sensitive and time critical targets:

> Time sensitive targets are those requiring immediate response because they pose (or will soon pose) a danger to friendly forces or are highly lucrative, fleeting targets of opportunity.[2]

> Time critical targets are time sensitive targets with an extremely limited time window of vulnerability, the attack of which is critical to ensure the successful execution of the Joint Task Force operations. They rank high on the joint integrated prioritized target list.[3]

These broad definitions of time sensitive and time critical targets include aircraft, cruise and ballistic missiles, air defenses, fielded ground forces, small naval craft, and diesel submarines, as well as lines of communication (such as roads, bridges, and rail lines) for movement of forces and high-value military cargo.

COMMAND AND CONTROL AND BATTLE MANAGEMENT

Both deliberate and dynamic C2BM practices are employed to monitor, assess, plan, and execute air and space operations. For example, the traditional air tasking order (ATO) cycle is predominantly deliberate (that is, targets are selected and weapons and targets are paired up to 72 hours in advance of the missions), but certain preplanned forces are designated to conduct dynamic

---

[2]AC2ISRC, 2000a, Sec 1.2.

[3]AC2ISRC, 2000a, Sec 1.2.

operations such as defensive and offensive counterair and close air support against emergent targets. DC2BM is therefore not new. What is new is the increased emphasis on less-traditional time sensitive/critical targets and enemy tactics that require the application of increasingly dynamic C2BM if these targets are to be successfully attacked. But key adversary center-of-gravity targets that provide critical support for military operations—such as industries and infrastructure, electric power, fuel production, storage and delivery systems, and lines of communications—will continue to be amenable to deliberate operations. In such operations, the targets are selected for attack to achieve the operational objectives of the overall air campaign plan. If operators overemphasize TCTs, air operations may degrade into a target-servicing process. Consequently, this study considered a balanced approach to integrating deliberate and dynamic C2BM.

ANALYTICAL APPROACH

We began our study with a review of selected recent operations (Desert Storm, Southern Watch, Northern Watch, and Allied Force) to identify key factors (such as political constraints, rules of engagement, and environmental characteristics) that shape DC2BM doctrine, concept of operations (CONOPS), and tactics, techniques, and procedures (TTP) and to identify missions that put a premium on dynamic operations.

We then examined new technologies and operational concepts that future adversaries may take advantage of to challenge our future DC2BM capabilities. The combination of shortfalls identified in recent operations and lessons learned by potential adversaries and their responses served to focus our examination on options to address those shortfalls.

Next, we reviewed ongoing and planned new programs and initiatives to improve U.S. DC2BM capabilities. We examined the Air Force's joint expeditionary force experiment (JEFX), wherein new processes, systems, and tools were tested to address C2BM shortfalls in time sensitive and time critical targeting. We also examined efforts to enhance DC2BM that are under way or planned by the services and joint communities, including new or improved sensor capabilities, evolving display technologies, collaborative environments, automated decision aids, expanded communications applications, and integrated assessment efforts.

Employing mission-level analysis, we then assessed the military value of candidate DC2BM enhancement options that included selected ongoing and planned programs and initiatives as well as those that we developed in the course of our mission analysis. To aid in this analysis, we decomposed DC2BM into six key functional areas, which we used as a framework to enable us to identify individual mission shortfalls. We then integrated the shortfalls across four mission areas. The six functional areas are as follows:

- Integrated tasking and rapid retasking of sensors and related processing, exploitation, and dissemination (PED).

- Timely integration (correlation and fusion) of information derived from multiple sources.

- Rapid target development and target nomination.

- Rapid weapon and target pairing.

- Timely decision and attack order dissemination.

- Rapid assessment of effects of weapon delivery.

The first two functional areas (which involve both ISR and C2BM tasks) are critical for developing an accurate intelligence preparation of the battlespace (IPB) and creating a common operational picture to help maintain accurate and timely situational awareness (SA) of friendly, enemy, and neutral air and ground assets. The effective direction, control, and management of assets (e.g., sensors, processing, analysis, and personnel) to develop IPB and SA are essential to support the commander's development of courses of action and the monitoring, assessment, planning, and execution of dynamic air operations.

Although "communications" does not appear explicitly in the names of any of the six functional areas, the ability to manage communications is inherent in all of them. Tactical communications within C2 nodes and between those nodes and combat aircraft is clearly an enabler of timely decision and attack order dissemination. In addition, on-demand, sufficient, and assured communications is one of the enablers of the robust, collaborative, distributed environment needed to accomplish these DC2BM functions.

Finally, taking into account planned investments in ISR and weapon systems, we identified suggested actions (both nonmateriel and technology-based) that should be considered by the Air Force and DoD to improve DC2BM of future air operations.

STRUCTURE OF THE REPORT

This report is an abridged version of a longer, limited-distribution report (Hura, McLeod, et al., 2002) that documents our analysis in much greater detail. In the present report, we focus on the DC2BM shortfalls identified in the detailed analysis and present suggested areas for improving DC2BM.

Chapter Two summarizes DC2BM shortfalls in four mission areas and provides suggested actions for improving DC2BM. The mission areas examined are counterair operations, theater missile defense, suppression of enemy air defenses, and interdiction.

Some DC2BM shortfalls are common across one or more mission areas, whereas others are mission-specific. In Chapter Three, we present a mission-based integrated assessment of DC2BM shortfalls and suggest actions for addressing these shortfalls. We use the six DC2BM functional areas discussed above to provide a framework for understanding this commonality and for integrating the shortfalls across mission areas. We also examine DC2BM-related infrastructure, organization, and CONOPS. Our top-level suggested actions and considerations are presented in Chapter Four.

MISSION-AREA C2BM ASSESSMENT

Our review and qualitative and quantitative analyses of past air operations and future projections identified those mission areas in which existing DC2BM capabilities are assessed to be adequate and those in which shortfalls exist. Not surprisingly, since we focused on dynamic operations, our analyses suggest that the performance of systems (C2BM, ISR, aircraft, and weapons) in each of the mission areas would benefit to some extent from improvements in the six identified DC2BM functions.

In this chapter, we summarize DC2BM shortfalls for four mission areas and suggest actions for improving DC2BM. The mission areas examined are (1) counterair operations, focusing on cruise missile threats; (2) theater missile defense (TMD), focusing on counterforce operations against transporter-erector-launchers (TELs); (3) suppression of enemy air defenses (SEAD) in the context of strike missions against targets defended by advanced air defenses; and (4) interdiction operations, focusing on small-unit ground forces intermingled with the civilian population.

COUNTERAIR OPERATIONS

The concept of dynamic command and control and battle management is not new to the U.S. Air Force: Counterair operations routinely rely on DC2BM. In fact, nowhere is DC2BM more critical than in the area of air-to-air engagements, where the speed of operations dictates close coordination among the various air defense elements in order to effectively, efficiently, and rapidly defeat the enemy air threat.

Along with SEAD, the Air Force conducts counterair operations to achieve and maintain air superiority and supremacy. Both offensive counterair (OCA) and defensive counterair (DCA) operations[1] are conducted to neutralize all enemy air threats, including fixed and rotary-wing manned combat and reconnaissance aircraft, unmanned aircraft (such as cruise missiles and unmanned aerial vehicles), and ballistic missiles.[2]

We examined the capabilities of the United States to defend against conventional military aircraft (fighters, bombers, and helicopters) by reviewing the results of past military operations (Desert Storm and Allied Force). Combat identification continues to be an issue, primarily because of limitations in sensor systems and identification friend or foe (IFF) systems. Such limitations can result in a requirement for visual confirmation of the air or ground[3] target by the shooter, which can put the shooter at greater risk. Except for such concerns, the U.S. Air Force is well-equipped and well-trained from a DC2BM perspective to conduct counterair operations against a conventional military aircraft threat posed by likely future adversaries in major theater wars and lesser-intensity conflicts.[4]

Counterair operations, however, may be less effective in the future because of the likely proliferation of cruise missiles. Cruise missiles are considered to be a low-cost solution for theater strike. Their small

---

[1]OCA operations include sweep, escort, and strike; DCA operations include combat air patrol and high-value air asset protection.

[2]Theater ballistic missiles are often treated separately from conventional aircraft and cruise missiles because of their unique characteristics and because of the effect they had on the conduct of the Persian Gulf War. We address them separately in the next section on theater missile defense because of their differences from cruise missiles and other aircraft. We recognize, however, that the Department of Defense (DoD) has a commitment to developing a joint theater air and missile defense (TAMD) capability that addresses all air threats. While there may be, a priori, common DC2BM elements in the preferred responses to the various future air threats, our approach has been to first treat each threat separately, from a more limited, focused perspective, to surface the most pressing DC2BM issues.

[3]For OCA strike missions, example ground targets include control towers; fuel storage areas; runways; and aircraft in the open, in hangers, or in hardened bunkers.

[4]DC2BM for counterair operations against enemy helicopters in a nonlinear battlefield is a challenging proposition requiring close coordination with the Army. This concept was not examined in the study. Also, asymmetrical threats such as hijacked commercial airliners by terrorists in suicide missions pose added DC2BM challenges that were not examined here.

size can make them difficult to defend against (if the missiles are air-launched, they also enhance the survivability of the manned aircraft that launch the missiles outside defended airspace). Further, the United States has demonstrated the military effectiveness of cruise missiles for theater strike. Finally, cruise missiles pose a significant new security challenge if armed with chemical, biological, or nuclear weapons of mass destruction (WMD).

Although existing DC2BM capabilities are adequate for counterair missions against conventional aircraft, our analyses suggest that successful DCA barrier operations against cruise missiles will require several enhancements to the current DC2BM process.[5] These enhancements would also improve operations against conventional military aircraft. Some of the enhancements are programmed or planned; others require further study. We grouped them into three categories: situational awareness, networked communications, and defense in depth.

Situational Awareness. Good situational awareness and an accurate air picture are needed by the Airborne Warning and Control System (AWACS) aircrew to support effective and efficient air-to-air engagements in a crowded airspace that contains coalition military aircraft (both conventional and stealthy) as well as enemy cruise missiles. These attributes are essential to ensure airspace control, deconfliction, and efficient allocation of interceptors to air targets.[6] The mission crew commander's guidance and direction are essential in building an accurate air picture and establishing good situational awareness. His guidance and direction establish cross-correlation requirements and identification thresholds to classify air contacts.

---

[5]A very advanced cruise missile threat would present a serious challenge to the Air Force. In addition to DC2BM enhancements, a new airborne sensor system with more power and a larger aperture (presumably, a phased-array antenna) would be needed to counter the missile's low radar cross section (Hura, McLeod, et al., 2001). An unmanned variant is also a possibility. Alternatively, a space-based system could be developed, if cost-effective. A new, manned airborne system would rely on DC2BM concepts discussed here; however, an unmanned airborne system or a space-based system, with the sensors separate from the C2BM functions, would require new concepts not addressed in this study.

[6]Good situational awareness and an accurate air picture will also support dynamic tasking (including route deconfliction) of fighters against ground TCTs.

Counterair operations normally require a rapid response to engage imminent air threats. This is particularly true for low-observable air targets such as cruise missiles; given the shorter detection ranges, there is less time to react. However, without an accurate air picture, commanders will rely on restrictive rules of engagement to minimize the chances of fratricide. Specifically, the AWACS mission crew commander or his designate will not have the necessary authority to declare air threats "hostile" and to direct airborne interceptors to engage them (i.e., to authorize weapons release). This was a major issue during Operation Allied Force.

Because of the importance of the counterair mission to the Air Force, that service should play a major role in developing capabilities for creating and maintaining an accurate air picture.[7]

An important attribute of an accurate air picture is that it should depict all air traffic in the theater of operations. As U.S. stealth assets and airborne reconnaissance assets such as unmanned aerial vehicles (UAVs) increase in number, it becomes more critical to find a method to incorporate them into the air picture. It is even more critical if these airborne reconnaissance assets deviate from their flight plans, for example, to collect information on a ground TCT. Civil air traffic and commercial air traffic can also be a source of confusion, especially if aircraft deviate from planned flight paths.

Increasing the detection range of AWACS will also improve situational awareness. The AWACS Radar System Improvement Program (RSIP) provides a factor-of-two improvement in detection range against low-observable targets;[8] complete implementation on AWACS is planned to be completed during 2005.[9] Another programmed improvement is the addition of Integrated Broadcast Ser-

---

[7]After this research was completed, the authors learned that senior DoD officials signed a charter on October 26, 2000, for a new joint-service task force that will identify the most effective and efficient means for building and maintaining a single integrated air picture that satisfies the warfighters' needs (Dupont, 2000). The Navy is the lead systems engineer, the Air Force is the deputy lead engineer, and the Army is the acquisition executive.

[8]ACC, 1996, p. 82.

[9]Air Combat Command announced initial operational capability (IOC) on June 14, 2001, with seven RSIP-equipped AWACS. Work on the remaining 25 AWACS is expected to be complete by February 2005 (Mayer, 2001).

vice receivers to ensure access to information collected by theater and national reconnaissance assets; receivers have been purchased and installation on AWACS has begun.

Upgrades to AWACS computers and displays would also improve the air picture. Specifically, a new tracker, improved IFF capabilities, and integration and fusion of onboard and offboard sensor data (e.g., multisensor integration) are needed to improve tracking and identification, and an improved graphical user interface technology (e.g., man-machine interface) is needed to reduce C2BM process timelines. The Air Force is planning to acquire these upgrades (under Block 40/45) when funding becomes available.

Networked Communications. Tactical digital information links (TADILs) offer network participants a number of advantages over voice communications. A wide range of combat data in addition to voice can be shared over a secure, jam-resistant communications network that is continuously and automatically updated. Data can be sent faster and more reliably via direct digital (i.e., computer-to-computer) communications, and text messages need only a small fraction of the communication resources that interactive voice messages require.

A number of TADILs have been developed to support near-real-time exchange of data among tactical data systems. The most recent is Link 16. Link 16 is an encrypted, jam-resistant, nodeless tactical digital data link network established by Joint Tactical Information Distribution System (JTIDS)–compatible communication terminals that transmit and receive data messages in the TADIL J message catalog.

In counterair operations, AWACS weapons directors must relay timely and accurate threat data to airborne interceptors, preferable via a tactical digital data link (such as Link 16) rather than secure voice communications. In addition, a data link will improve coordination among the participants of the counterair operations. In particular, situational awareness would be enhanced among network participants by the network's automatic and continuous update of friendly tracks as well as tracks of hostile aircraft.

AWACS, Rivet Joint, ground C2 centers, and one squadron of F-15Cs already are equipped with Link 16 terminals (specifically, JTIDS terminals). The Air Force plans to equip the rest of its fighters (F-16s,

F-15C/Es) with Link 16 terminals (specifically, Multifunctional Information Distribution System terminals).

Other networked communications may be needed to support defense in depth, as discussed in the next section.

Defense in Depth. Although not explicitly examined because our focus was on DCA barrier operations, we believe that a multilayer defense construct would improve defense effectiveness against cruise missiles and is probably needed against stealthy variants. The Air Force should develop DC2BM CONOPS and capabilities for engaging cruise missiles and other stealthy aircraft that include defense in depth.[10]

In particular, procedures and systems for disseminating guidance and orders and sharing of sensor data and information from one layer to the next are essential. Thus, fighters conducting OCA sweeps (e.g., F-22s) should relay cruise missile detections and results of engagements to AWACS supporting DCA barrier operations. Currently, F-22s are planned to have a receive-only Link 16 capability when the F-22 program reaches initial operating capability in December 2005, to support their OCA operations. This is perceived as a shortfall for defense-in-depth operations (F-22 could use secure voice communications to provide some situational awareness back to AWACS).[11]

Similarly, AWACS should relay data and directions (if assigned authority) on cruise missile leakers of barrier operations to terminal area and point air defenses. This could occur via a joint data network (JDN) such as Link 16 (both the Army and Navy have or are acquiring Link 16 capabilities), or via a joint composite tracking network (JCTN) such as the Navy's Cooperative Engagement Capability

---

[10]Another defense-in-depth construct, one for theater ballistic missiles, is discussed in the next section. The defense-in-depth construct for engaging aircraft and cruise missiles should be combined with that for theater ballistic missiles to create an overall joint theater air and missile defense construct.

[11]After this research was completed, we learned that the Air Force believes that it will be able to field Link 16 transmit capability for the F-22 in about 2007 in the aircraft's first post-IOC software upgrade (Wolfe, 2001).

(CEC).[12] In all likelihood, both will be used by terminal defenses. Thus, interfaces between JCTN and JDN will be needed.[13]

THEATER MISSILE DEFENSE

Allied experience with Iraqi mobile Scud theater ballistic missiles (TBMs) has led to increasing emphasis on systems and operational concepts for dealing with mobile, time critical targets of this sort. In fact, for a number of years, the Scud's TEL[14] was the prototypical TCT.

Shortfalls in ISR, C2BM, and weapon systems contributed to the Allies' poor performance against TBMs during Operation Desert Storm. Since then, DoD has responded with a number of initiatives to improve U.S. capabilities against this threat. Because no single defense element can provide the desired level of performance (near-zero leakage), especially against TBMs armed with WMD, there has been substantial emphasis on developing a multilayer defense capability (i.e., defense in depth) for TMD.[15]

_____

[12]JDN and JCTN are two of the three communication network constructs defined by the Joint Theater Air and Missile Defense Organization (JTAMDO) (the third is the joint planning network [JPN]). A JDN is a collection of near-real-time communications and information systems that permits ground, air, and sea controllers to provide (and exchange) ISR and C2 information to shooters. A JDN carries near-real-time tracks, unit status information, engagement status and coordination data, and force orders. Among other data, it provides the identification, location, heading, speed, altitude, and status of friendly aircraft, missiles, ships, submarines, and selected ground systems; and the location, heading, speed, altitude, and (if available) the identification of neutral and enemy air, sea, submarine, and ground contacts. A JCTN is a real-time sensor fusion system that enables ships, aircraft, and ground air defense systems to exchange sensor measurement data to create common composite air tracks of fire control accuracy. It includes common software and a communication element that allow participating units to share and fuse real-time sensor data.

[13]This is not a trivial task; the Navy has experienced problems integrating Link 16 and CEC on its Aegis ships.

[14]More specifically, the Scud TEL was the Soviet MAZ 543.

[15]Defense in depth provides multiple shot opportunities against difficult targets such as TBMs, and is a recognized concept within the defense community. It is one of the four basic operational objectives for a joint theater air and missile defense capability. The other three are single integrated air picture; early detection, classification, and identification; and 360-degree coverage (JTAMDO and BMDO, 1998, pp. 2-10 to 2-13).

In our analysis, we focused on Air Force counterforce operations against the TELs—both prelaunch and postlaunch operations—and the necessary DC2BM to support these operations. Counterforce operations, to the extent that they can be successful, clearly benefit the other elements of the TMD defense-in-depth architecture by reducing the number of missiles and the salvo potential that other defense elements (boost phase, midcourse, terminal, and passive defenses) will have to counter. If the enemy's TELs can be systematically destroyed, fewer missiles will be launched, resulting in fewer missiles to be countered by the next defense layers and hence in less damage to friendly forces and population. In the limit, if the number of TELs is small compared with the number of reload missiles, killing all the TELs, even postlaunch, can ground reloads. Given the current expense and complexity of TELs, it would be expected that the number of reloads per TEL might be quite high.

Our analysis shows that current DC2BM capabilities are inadequate to support prelaunch and postlaunch counterforce operations against TBM TELs. Improvements in all six DC2BM functional areas are essential to ensure that the DC2BM process is completed in the tight, threat-driven timelines—with timelines approaching 10 minutes or less, measured from the time of initial detection of a possible TCT to order issuance to a strike aircraft.[16] Moreover, TMD-related DC2BM initiatives should initially focus on postlaunch counterforce operations since this mission is best supported by existing ISR capabilities.

We use the following four categories for our discussion of potential improvements: new sensors and sensor upgrades, fighter sensor upgrades, dynamic C2BM processes and tools, and collaborative environment and networked communications.

New Sensors and Sensor Upgrades. As with most TCT challenges, solutions to TMD counterforce problems will require significant improvements in our ISR capabilities. Because of the depth and operational flexibility inherent in long-range TBM threats, the United States will require new sensors and sensor platforms. These must address the serious challenges of providing affordable and militarily

---

[16]This timeline does not include the flight times of retasked sensor(s) and strike aircraft to possible targets.

useful capabilities, including deep-look, long-dwell, all-weather/day-night operations, and acceptable survivability in the face of advanced air defenses. For postlaunch operations, focused-look surveillance capabilities are sufficient, whereas for prelaunch counterforce operations, broad-area search with much higher search rates at higher imagery resolution (relative to existing capabilities) is needed. These advanced capabilities will be needed in peacetime, day-to-day, for intelligence and strategic warning; in short-notice crises; and in war.

Upgrading existing sensor systems also offers promise for scenarios in which standoff sensor platforms can surveil the deployment areas or in which air defenses have been sufficiently suppressed to allow increased penetration by sensor platforms. For example, planned and proposed upgrades for Joint Surveillance [and] Target Attack Radar System (JSTARS) should improve the platform's ability to detect, identify, and track stationary and moving vehicles. The upgrades include (1) integration of the advanced radar being developed under the Multi-Platform Radar Technology Insertion Program (MP-RTIP), which will significantly shorten the revisit rate over the ground radar coverage area; (2) development of better tracking algorithms to improve the chances that identified targets will be continuously tracked until an attack asset arrives; (3) development of better ground moving target indication (GMTI) discrimination capabilities beyond the double-Doppler wheeled/tracked distinctions—for example, a high range resolution (HRR) GMTI radar mode and associated algorithms that can exploit the collected range profiles; (4) development of an all-source integration capability to aid in understanding TBM/TEL operations and identifying targets—for example, the target evidence accumulator; (5) development of improved imagery geolocation capabilities—for example, JSTARS Imagery Geolocation Improvement (JIGI);[17] and (6) integration of appropriate TADIL J messages to support ground attack (i.e., JSTARS Attack Support Upgrade program). Many of these capabilities could also be applied to alternative aircraft platforms, such as high-altitude en-

_____

[17]JIGI is an Air Force Tactical Exploitation of National Capabilities initiative. It is a stand-alone workstation that correlates JSTARS data with geocoded archival national imagery to improve the geolocation accuracy of detected tactical targets.

durance UAVs (e.g., Global Hawk) or a new multimission C2ISR aircraft.[18]

While such sensor and data exploitation upgrades will improve the capabilities of existing manned platforms to support missions such as TMD counterforce, the high-value platforms themselves have deficiencies (e.g., in survivability, access, responsiveness, operations tempo) that may limit their effectiveness for these challenging counterforce missions, even with improved sensor and processing systems. Consequently, it is important to build in as much adaptability and scalability as possible in these sensor advances to support their future application in other platforms (e.g., stealthy UAVs and satellites).

Fighter Sensor Upgrades. Sensor upgrades may also be appropriate for the fighters, as well, depending on the specific operational concept (e.g., the extent to which the fighters can rely on offboard targeting [for instance, receiving coordinates for Joint Direct Attack Munition employment] versus having to find, track, identify, and target the TELs themselves based on relatively crude offboard cues). For example, improving the resolution of the F-15E's APG-70 in the synthetic aperture radar (SAR) mode, adding a GMTI capability, providing the ability in the air-to-air mode to track (and, hence, backtrack) inflight ballistic missiles, and sensor management aids to improve area search may all be worthwhile improvements for the TMD counterforce mission (and are likely to be possible with relatively inexpensive software modifications). Further in the future, initiatives such as the Advanced Targeting Pod may be critical in improving the shooters' ability to engage these targets under a variety of challenging operational conditions.

Dynamic C2BM Processes and Tools. Improvements in C2BM capabilities are also needed. For example, decision aids, based on

---

[18]During the course of our analysis, the commander of Air Combat Command published a proposal advocating a multimission aircraft (MMA) that would be used to support the Expeditionary Air Force more fully, incorporating capabilities of AWACS, JSTARS, Rivet Joint, and the Airborne Battlefield Command and Control Center (ABCCC), and adding an airborne information fusion functionality (Jumper, 2001). In addition, the AC2ISRC is developing an operational concept for multimission C2ISR aircraft (AC2ISRC, 2000d). More recently, the AC2ISRC concept has furthered evolved into the multisensor command and control aircraft (MC2A) (AC2ISRC, 2001).

game-theoretic analysis that explicitly evaluates the outcomes of two-sided games with alternative Blue and Red strategies,[19] may be valuable in supporting responsive, rationalized force allocation decisions by combat plans staffs. As U.S. TMD capabilities are developed and fielded, potential adversaries will respond. The analysis of these action/reaction cycles (using, for example, simple tools from game theory) is critical to determine the most profitable strategies for the United States to adopt. An effective TBM defense posture will most likely involve defense in depth, which includes prelaunch and postlaunch counterforce operations and active and passive postlaunch defenses.

Combat operations will also need enhancements to support prelaunch and postlaunch counterforce operations. The result of these enhancements should be measured in terms of very short C2BM timelines (approaching 10 minutes or less). The Air Force should work to achieve the technical, cultural, and organizational changes that will make this improvement possible. Perhaps most important is the development of a TCT functionality that includes TTP, personnel, and systems for (1) developing a good IPB, (2) maintaining an accurate, near-real-time, integrated air and ground situational awareness; (3) performing rapid target development and nomination; (4) performing rapid weapon and target pairing; (5) ensuring rapid decision and order issuance; and (6) ensuring timely dissemination of information necessary for mission execution and assessment.

The development of necessary TCT functionality in the short term should concentrate on the postlaunch counterforce problem, where improvements are likely to come most readily, but should also include elements that will facilitate the integration of all phases of TMD operations.

Collaborative Environment and Networked Communications. To accomplish these functions in times approaching 10 minutes or less, a robust collaborative environment is needed that includes automated tools; on-demand, high-data-rate communications; a robust network and server architecture with responsive operating protocols;

---

[19]Examples of such analyses are given in Hamilton and Mesic, 2001, and Hura, McLeod, et al., 2002, Appendix B.

an expert network manager; and an expert and empowered information manager. Further extensive automation of data management and applications to assist operators perform the DC2BM functions is essential.

SUPPRESSION OF ENEMY AIR DEFENSES

SEAD is a critical mission area because its success or failure has a significant impact on all air operations and supported ground operations. SEAD has two objectives: to minimize friendly aircraft attrition and to maximize air power flexibility and effectiveness. Threat avoidance is the preferred doctrinal means for minimizing aircraft attrition. This is usually accomplished by route planning, stealth, and standoff weapons. In those instances when threat avoidance is not a viable option (e.g., when threat locations are poorly known) or the demands of the military effort require the threat to be overcome to maximize air power flexibility, effectiveness, and support to ground operations, suppression or destruction of the threat is necessary.

Military operations during the 1990s clearly showed that U.S. air forces can help win the war while concurrently maintaining very low levels of aircraft losses in the face of most currently deployed air defenses of likely adversaries. The combination of threat avoidance, judicious tactics, self-defense capabilities, and suppression packages (e.g., jamming aircraft, aircraft carrying anti-radiation missiles, air-launched decoys), and associated DC2BM has been effective against older generation surface-to-air missiles (SAMs).

However, these successes have not been without considerable costs and risks. SEAD requires significant allocation and application of forces (aircraft and people), increases organizational strain, and limits preferred attack CONOPS. Low-level air operations, in particular, have been hampered by an inadequate ability to detect, locate, identify, track, and target anti-aircraft artillery and man-portable infrared-guided and electro-optical-guided SAMs and anti-aircraft artillery. This has been a persistent ISR shortfall and we do not offer an ISR solution here. Existing workarounds to deal with these threats suggest a continued focus on DC2BM efforts that support threat avoidance, judicious tactics, and self-defense measures.

Successful air operations are expected to be further challenged if future adversaries deploy more-capable air defense systems, such as the SA-10 and SA-20. The worrisome trends associated with these emergent threats include increased range, lethality, mobility, and netting (i.e., development of an integrated air defense system).

The expected performance of these advanced air defense systems places current CONOPS and weapon systems at increased risk and will require improvements in all aspects of the engagement process. Responses to such emergent threats will likely involve a mixture of improvements to ISR systems, C2BM capabilities, weapon systems, and communications. We use the following four categories for our discussion of potential improvements: situational awareness and new sensors, dynamic C2BM processes and tools, new weapons, and networked communications.

Situational Awareness and New Sensors. Situation awareness in an environment with advanced SAM systems is critical. It must be timely and accurate, providing knowledge of threat location, tactics, and strategy to both strike aircraft and SEAD platforms. Improved signals intelligence systems are needed that can operate across the entire signal spectrum and that can quickly identify and accurately locate emitters that operate intermittently. There will be increased emphasis on, and requirements for, ISR sensors that can surveil mobile targets until weapon systems achieve target destruction. Specifically, there is a need for focused-look, long-dwell imaging sensors capable of operating day and night in unfavorable weather conditions while surviving in a severe threat environment. Without these capabilities, U.S. military forces cannot effectively detect, locate, identify, track, target, and engage advanced SAMs.

Dynamic C2BM Processes and Tools. More-dynamic C2BM systems, procedures, and CONOPS are needed to shorten response times so that mobile SAMs can be killed before they flee. Automated tools and procedures are needed for rapid retasking of sensors and associated PED, and for processing and correlation of information from multiple intelligence disciplines (i.e., multi-INT fusion). Decision aids are also needed to more effectively support the dynamic management of future integrated strike and SEAD packages. These aids should provide needed functionality in the air (e.g., air command element, C2ISR package commander[20]) and on the ground (e.g., TCT cell or combat operations division in the air operations center [AOC]). SEAD CONOPS must be updated to reflect the need for more-dynamic C2BM procedures. They must also be adapted to incorporate not only future kinetic (explosive) weapon systems but also new nonkinetic information warfare (IW) techniques. Furthermore, these CONOPS must be flexible enough to respond appropriately to variations in enemy air defense strategies ranging from maximum engagement and concomitant friendly-force exposure to episodic engagements with minimal exposure.

Automation can also enhance the Air Force's ability to dynamically reroute strike aircraft, which should greatly improve strike aircraft survivability and SEAD effectiveness. Although perfect situational awareness is always preferred, the consequences of imprecise, inaccurate, or obsolete information resulting from mobile air defenses can be minimized when inflight updates to strike platforms—including rerouting information—enable aircraft to avoid encounters or at least minimize their effects. Supported by appropriate communications, autorouting tools will also enable strike aircraft to respond to the dynamics of SEAD operations.[21]

New Weapons. Although necessary, the improved ISR and C2BM capabilities discussed above are not sufficient. SEAD weapon systems also need increased stealth, range, and lethality. The fielding of stealth aircraft and longer-range standoff weapons allows the Air Force to strike targets with minimum exposure time in the lethal range of enemy air defense systems. But these weapons must be cued. Cueing calls for enhanced DC2BM capabilities that support

---

[20]The C2ISR package commander is a new concept that is an outgrowth of identified needs in exercise, experiment, and operational situations. A formal C2ISR package commander course is being taught at the AWACS wing and was showcased for the first time during the Red Flag exercise held at Nellis AFB during January–February 2001. The C2ISR package commander would be responsible for orchestrating all of the C2, ISR, and electronic support assets executing in the joint force air component commander's tactical area of responsibility (552nd ACW, 2000).

[21]Note that autorouting tool development will expand flex targeting opportunities for targets other than air defenses. Allied Force flex targeting was limited in its success, and one of the reasons was the inability of most of the force to be dynamically rerouted and mission planned.

rapid retargeting of new standoff weapons. In addition to kinetic weapons, electronic warfare and innovative IW methods will likely be required to combat netted air defense systems.

Networked Communications. Finally, to better support current SEAD CONOPS and enable future integrated strike/SEAD CONOPS in the presence of advanced air defenses, development of networked communications should be continued. Current programs for improving communications, such as Link 16, Improved Data Modem,[22] and various gateways,[23] should be synchronized (through interoperable systems or via gateways), and new capabilities fielded as technologies allow.

INTERDICTION

Air interdiction or battlefield air interdiction, herein simply "interdiction," is performed by the Air Force to help shape the battlefield, to attrit hostile forces, and to disrupt enemy lines of communication and logistics support. Typically, interdiction missions have been flown in support of joint air-land campaigns. Operation Desert Storm and subsequent interdiction campaigns, however, have deviated somewhat from this combined-arms model.

Large Armor Formations

Interdiction during the halt phase of a major theater war remains a critical military mission to minimize the territory captured by an invading army and to protect forces currently in theater. The mission is particularly challenging because the targets are mobile, hard to kill, and defended by fixed and mobile SAMs and anti-aircraft artillery. It

---

[22]Improved Data Modem is a high-speed digital data link modem that is an interface between the aircraft's onboard radios (HF, UHF, VHF, or satellite communications) and its mission computer. It can pass near-real-time mission and targeting data among ground-based observers, military aircraft, attack teams, and artillery fire direction centers. The system is fielded on select Air Force aircraft, such as Rivet Joint, JSTARS, and F-16 Block 40 and Block 50. It is also fielded with tactical air control parties and at air support operations centers.

[23]*Gateways* are devices that support information exchange between a variety of communication systems to ensure the rapid exchange of information among C2BM nodes and between these nodes and attack assets.

becomes even more challenging in a short-warning scenario that quickly leads to open hostilities and necessitates the rapid deployment of additional forces to theater.

Recent RAND analyses[24] of a postulated Southwest Asia scenario indicate that current and planned programs (sensors, C2BM, and weapons) will maintain the effectiveness of interdiction assets against ground forces in large formations. However, improvements in two areas (sensor and battle management upgrades and networked communications) could improve efficiency.

Sensor and Battle Management Upgrades. A number of planned and proposed JSTARS upgrades—MP-RTIP, a better tracking algorithm (e.g., kinematic automatic tracker), better GMTI discrimination capability (e.g., HRR), a multisource integration/fusion capability (e.g., target evidence accumulator), imagery geolocation improvements (e.g., JIGI), and updates to the TADIL J message set (i.e., the Attack Support Upgrade program)—should increase efficiency of interdiction operations.[25] Another useful improvement is a decision aid to support targeting, especially for standoff weapons; this tool would rely on the information provided by many of the above initiatives.

As discussed in the previous sections, these improvements have broad applicability to other airborne platforms (e.g., UAVs) and across a number of mission areas. In fact, they are critical capabilities for TMD and SEAD. Moreover, these improvements should be developed so that they can be integrated into ground C2BM facilities that control future UAVs and space-based radars.

Networked Communications. Finally, in the near term, networked communications systems are expected to provide connectivity for ground and airborne C2BM nodes and strike aircraft. In particular, Link 16–coordinated operations could significantly improve the flow of strike assets in situations where large formations of armor are attacking.

---

[24]See Ochmanek, et al., 1998; and Hura, McLeod, et al., 2000, pp. 150–173.

[25]These JSTARS upgrades also apply to TMD and were described in more detail in the TMD section of this chapter.

Small-Unit Ground Forces

Traditionally, the Air Force has focused on the interdiction of large armor formations as a key mission and has devoted less attention to air-to-ground operations against small-unit ground forces. Over the past ten years, however, it has attempted to use air power against such forces. Air operations against small-unit ground forces are particularly difficult when a corresponding friendly ground component is not available to help detect, track, and identify targets on the ground, and then to assess the results of air-to-ground engagements. As demonstrated in Operation Allied Force and other recent operations, C2BM for the interdiction of small-unit ground forces has been poor for a number of reasons, including difficulties with positive identification of potential targets and with collateral damage assessment. Improvements in three areas (situational awareness, decision aids, and joint CONOPS) could help remedy deficiencies.

Situational Awareness. An integrated air and ground picture should be developed jointly by the services. Such a solution would support the need for positive identification of forces that are intermingled with the civilian population. Because this function is performed primarily with ground sensors and may be much more difficult with air or space sensors, involvement by the Army and special operations forces is essential (however, in certain hostile environments, their employment may not be worth the risk). Making such an integrated air and ground picture timelier requires automated tools to rapidly retask surveillance and identification sensors and provide associated PED.

Decision Aids. Decision tools are needed to help commanders assess the situation, determine collateral damage, recommend response options, and support a go/no go decision.

Joint CONOPS. The Air Force should support improved coordination of the joint force land component commander and joint force air component commander. This can be accomplished through a review and modification of doctrine, CONOPS, and TTP, reinforced through practice and joint exercises.

On the latter issue, our research suggests that developing an operational concept (including TTP, ISR capabilities, and weapon systems) having characteristics comparable to police force operations offers a reasonable approach for addressing this very difficult mission, at least in low-threat environments or in hostile environments where the benefit outweighs the risk. Such operations typically consist of (1) rationalized surveillance[26] by remote and in-place sensors (video cameras in banks; foot, car, and helicopter patrols), supplemented by the observation of interested parties (local populace, friendly forces), to detect and report the occurrence of an undesired event (burning of a village); (2) a rapid multiphased response (dismounted and mounted ground and air elements) to contain perpetrators within a small area; (3) high-resolution, small-area sensors to distinguish perpetrators from local populace and friendly forces; and (4) lethal ground and air elements to positively identify and capture or, if necessary, eliminate perpetrators.[27] To implement such capabilities in military operations requires the close coupling of airborne and ground controllers and decisionmakers in the AOC (as well as ground C2BM nodes of the other services) with air and ground force elements, and automated tools to rapidly perform all six of the DC2BM functional areas.

_____

[26]Here, *rationalized surveillance* means the optimal allocation of scarce surveillance resources.

[27]While this concept has been offered as one option for further investigation, differences between the operational situations faced by the police and the military will, in all likelihood, affect the military's specific implementation and rules of engagement. The nature of the perpetrators is one area of differences. Usually, suspects in police operations are not heavily armed and are interested in escape and evasion (they have already committed the crime) rather than confrontation. Moreover, the police are interested in apprehending suspects for later prosecution, whereas the military may have other more immediate and lethal objectives.

Chapter Three

CROSS-MISSION DC2BM ASSESSMENT

In the preceding chapter, we summarized our assessment of DC2BM for four mission areas. Some DC2BM shortfalls are common across one or more missions; others are mission-specific. In this chapter, we present a mission-based integrated assessment of DC2BM shortfalls and suggest actions for addressing these shortfalls.[1] We use the six DC2BM functional areas presented in Chapter One to provide a framework for understanding this commonality and for integrating the shortfalls across missions. We also examine DC2BM-related infrastructure, organizations, and CONOPS.

As discussed in the mission area assessments, addressing the DC2BM shortfalls is not sufficient to achieve mission success against TCTs. In some cases, new sensor capabilities and weapons are also required. Moreover, large investments are being made to field new ISR assets, aircraft, and weapons,[2] and the proposed DC2BM enhancements must be considered in light of these investments.

The Air Force has recognized most of the shortfalls discussed in Chapter Two and is pursuing programs and new initiatives to address some of them. We reviewed TCT-related documents from the

---

[1]While much of the discussion in this chapter is applicable to airborne targets, the focus is really on ground TCTs encountered in the TMD, SEAD, and interdiction missions. Many shortfalls for airborne TCTs are mission-specific and are identified in the counterair mission in Chapter Two.

[2]These investments range from the billions to the tens of billions of dollars for individual programs, according to cost data from Selected Acquisition Reports. In contrast, C2BM programs are relatively modest, with most of them below $100 million, with total C2BM investment on the order of $2 billion (AC2ISRC, 2000b).

services, the unified commands, the Office of the Secretary of Defense, and national agencies;[3] observed or reviewed experiments (e.g., Joint Expeditionary Force Experiment, Millennium Challenge, Attack Operations Against Time Critical Targets, Roving Sands 2000); and discussed programs with subject-matter experts. After reviewing these efforts to address TCT challenges, we focused on two major Air Force efforts—the "Top 30" recommended solutions from ACC[4] and the initiatives in JEFX 2000[5]—because they appear to encompass most of the functional DC2BM needs identified in our mission area analyses. Our review of systems and capabilities was accompanied by an assessment of the processes those systems were designed to support and how those processes (existing and proposed) would also benefit from the application of these new capabilities.

We then mapped selected programs and initiatives from these two Air Force efforts onto the shortfalls identified in the mission assessments and determined remaining gaps in DC2BM capability.[6] Finally, we defined suggested actions to address the identified gaps, taking into consideration existing and new ISR and weapon system programs.

For each of the six DC2BM functional areas, Table 3.1 lists sub-areas that need improvement. The table also lists sub-areas requiring improvement in two other DC2BM-related areas: infrastructure, and organization and CONOPS.[7] In the remainder of this chapter, we review these eight sub-areas in more detail.

---

[3]AC2ISRC, 2000a; AC2ISRC/AFC2TIG, 2000a, 2000b; AC2ISRC/C2N, 2000a, 2000b; AC2ISRC/JWID JPO, 2000a, 2000b; ACC, 1997, 1999, 2000c; Boyle, 1999; JC2ISR JT&E, 2000; USAF/XOI, 2000; USJFCOM/J9, 1999, 2000a, 2000b.

[4]AC2ISRC, 2000b.

[5]AC2ISRC/AFEO, 2000a; AC2ISRC/C2B, 2000a, 2000b.

[6]Sauer, et al., 2000.

[7]This report only summarizes capability shortfalls and does not specify system shortfalls. Such a discussion can be found in Hura, McLeod, et al., 2002.

Table 1

DC2BM Areas for Further Improvement

DC2BM Functional Area, Infrastructure, CONOPS, and TTP		Sub-Areas Needing Improvement
Functional area	Sensor/PED retasking	Integrated multi-INT cross-domain tasking linked to guidance and mission objectives
	Correlation/ fusion	Integration of multisource (multi-INT and cross-domain) data (including that of combat platforms) and information to create common operational and tactical pictures; duplicate displays with common colors and icons
	Target development/ nomination	Automated nomination tool that considers restricted target list; tool to help track TCT nomination relative to ATO and JIPTL; integration of nonkinetic (e.g., IW) means
	Weapon-target pairing	Automated tool to help perform rapid weapons-target pairing for all weapon systems
	Order issuance	Standardized dynamic battle order (4-line or 9-line reports); decision aids for airborne command element and strike control element; combat platform autorouters
	Assessment	Automation-assisted assessment process; shooter data backlinks
Infrastructure, organization, and CONOPS	Infrastructure	Collaborative environment with voice, data, and multimedia capabilities; on-demand assured communications; robust network-server architecture; integrated systems and processes across all levels; information manager
	Organization and CONOPS	Integrated C2BM of ISR, shooters, operations, and combat assessment

NOTE: Acronyms are defined in the front matter of this report.

INTEGRATED TASKING AND RAPID RETASKING OF SENSORS AND PED

A major shortfall to optimal allocation of scarce ISR resources and to cross-cued and/or simultaneous collections is the lack of agreed-upon (by DoD and the intelligence community) CONOPS, TTP, and automated tools for integrated tasking and battle management of (1) sensors from multiple intelligence disciplines (multi-INT),[8] (2) sensors from multiple platform domains (cross-domain),[9] and (3) the associated PED to support military monitoring, assessment, planning, and execution processes and timelines. The *Intelligence, Surveillance, and Reconnaissance Integrated Capstone Strategic Plan*[10] underscores the need for these capabilities.[11]

The present process for tasking and retasking of sensors relies on tools developed for individual intelligence discipline (INT) sensors; usually, the tools also depend on the sensor platform. Even for conducting cross-domain, multi-INT tasking simply to allocate the best sensor (regardless of domain or INT) to a target, the current collection management process is manual and time-consuming.[12] If cross-domain, multi-INT tasking is desired for more-sophisticated collection operations (i.e., cross-cueing and simultaneous collections), further time-consuming manual interventions—mostly by personnel remote from the AOC—are necessary. Moreover, once tasking is completed, there are no management tools for PED tasking to ensure that collected information is processed and analyzed to support user timelines.[13]

---

[8]The four principal intelligence disciplines are imagery intelligence, signals intelligence, measurement and signature intelligence, and human intelligence.

[9]*Cross-domain* refers to integration across platform operating domains (e.g., space, air, land, and sea).

[10]ASD(C3I), 2000a.

[11]These issues are also highlighted in ASD(C3I), 2000b, and in OASD(C3I)/DASD(C3ISR&S), Joint Staff/J2, and Joint Staff/J6, 2000.

[12]For example, the process consumed a large fraction of a 24-hour day during Operation Allied Force.

[13]Although we have used the term "tasking," we are really speaking about collection management. For theater and tactical assets, the sensors can be tasked by theater collection managers. As discussed earlier, space-based ISR sensors are controlled and tasked by national intelligence agencies, and theater collection "requests" are for-

In addition, the current PED tasking and retasking processes are not well-integrated with the monitoring, assessment, planning, and execution processes and timelines of weapon systems at various levels of command (e.g., unit, wing, air operations center, joint task force). Typically, the outputs from a single-INT PED are geared more toward the support of specific standing requests for information or general updates to an overall geographical area of interest rather than to the support of highly dynamic operations. The lack of adequate rapid integration of cross-domain multi-INT PED outputs (products and services) and lack of rapid delivery of these outputs to the various levels of command often impede the rapid development of alternative courses of action and the execution of counter-TCT operations.

Efforts to better leverage and integrate scarce intelligence production and exploitation resources (analysts and systems) and targeting resources (targeteers and systems) are partially addressing the PED and force application integration problem. DoD has begun to employ federated intelligence production and exploitation processes and federated and collaborative targeting processes; both processes were recently employed in Allied Force.[14] Such processes must be standardized, refined, streamlined, more automated, and integrated to provide more responsive support to TCT operations.

The most important action that should be taken in this functional area is the development of an agreed-upon (by DoD and the intelligence community) CONOPS and TTP for integrated cross-domain multi-INT tasking and rapid retasking of sensors and PED, with PED products and services linked to military monitoring, assessment, planning, and execution processes and timelines. Without this action, process and tool development efforts will remain fragmented.

---

warded to these agencies. However, if military operations are under way, the theater collection manager can count on a significant fraction of the day's collection.

[14]The participants in these federated and collaborative processes were at widely geographically distributed centers; they primarily analyzed fixed targets. In some cases, however, there was more-direct support to time critical targets. For example, imagery analysts at the unified command's joint intelligence center would alert the combined air operations center (CAOC) when they detected and identified elements of Serbian air defense systems.

TIMELY INTEGRATION OF INFORMATION

For the purpose of this discussion, we define this DC2BM function as the timely control and management of

- access to multisource and multi-INT data and information

- correlation of data and information into data streams, information products, and services

- display of these diverse forms of data and information to support military users.

The actions suggested in the preceding section and those subsequently discussed in the infrastructure section largely address data and information access control and management issues. Therefore, this section focuses on correlation and display issues.

Currently, extensive ad hoc manual interventions are necessary to correlate and display cross-domain multi-INT data and information for use in the monitoring, assessment, planning, and execution of military operations. These interventions are necessary because of the following:

- Data are collected by multiple, often stovepiped, sensor systems that have diverse phenomenology (e.g., electro-optical, infrared, SAR, GMTI, acoustic).

- Different organizations and activities are responsible for transforming data and information into products and services to support military operations. Moreover, they may apply different data correlation tools, techniques, or algorithms to derive products and services.[15]

- Data streams, products, or services may not include accurate time stamps, and their spatial reference frames and accuracy may vary.

---

[15]For example, electronic intelligence data may be correlated at the pulse level or at a data product level. Similarly, radar data may be correlated at the plot level (unfiltered signal information captured by the radar receiver), or track level (filtered information based on a track initiation threshold setting—e.g., three out of five signal returns within the preset range and speed gate).

- Numerous displays have emerged to assist decisionmakers visualize correlated information.

In Operation Allied Force, CAOC ISR cell personnel and combat operations personnel had direct real-time feeds and displays of data from individual sensors (e.g., Predator, JSTARS), as well as intelligence products and services from other organizations (the unified command's joint intelligence center and CONUS-based exploitation centers). However, they had to manually integrate all this information to prosecute ground TCTs.

As the Air Force moves into using multiple GMTI information sources (e.g., JSTARS, Global Hawk, space-based radar[16]), it must consider how best to integrate the data from these sensors and convert the data to useful information on moving ground targets of interest to users. Also important will be the development of techniques and procedures for integrating GMTI and video information and other sources of data (e.g., hyperspectral imagery) that can help locate, identify, and track moving TCTs.

The emergence of fighter and bomber aircraft (e.g., F-22, Joint Strike Fighter) with improved sensors and improved data processors presents a further opportunity to improve situational awareness across all echelons of command. However, to realize the potential of these new sensing capabilities, data from fighters and bombers have to be backlinked to data correlation nodes (e.g., the control and reporting center) and integrated with relevant data from other sources (e.g., national reconnaissance and airborne surveillance sensors). Operational and technical architectures for such data collection, correlation, and information dissemination must be defined, agreed upon, and funded.

---

[16]Recently, DoD has taken a renewed interest in space-based radar concepts (DoD, 2001). Much of the discussion appears to focus on the space components and associated technology. Although these areas clearly need addressing, another critical element for an operational system is an agreed-upon CONOPS. The CONOPS should address a number of C2BM issues (both deliberate and dynamic), such as the relationship between the theater C2BM nodes and the satellites' payload control center; the C2BM process for sensor tasking, data fusion, and information dissemination; and the level of integration and interoperability with airborne SAR and GMTI assets (both U.S. assets and the assets of its allies).

The preceding improvements and efforts indicate that primary elements for creating a single integrated ground picture are being fielded. However, TTP and systems to integrate already available air and ground contact data have not yet proved effective. Increased emphasis on joint collaboration is needed to address this important deficiency. A single integrated air and ground picture of the battlespace across tactical and operational levels is required to support TCT operations.[17]

RAPID TARGET DEVELOPMENT AND TARGET NOMINATION

DoD has been placing increased emphasis on federated and collaborative targeting processes, but for the most part the processes are nonstandard, employ limited automation, and address primarily fixed targets. The current process for rapid target development and nomination relies primarily on in-house "sneaker networks"[18] to pass information among preassigned or selected ad hoc personnel who determine the location, identity, and criticality of a TCT and check it against targets that are already on the joint integrated prioritized target list (JIPTL) and that have been assigned for attack on the ATO.

A key capability, not only for rapid target development and nomination but also for weapon and target pairing and attack order development (i.e., the next two DC2BM functional areas), is to follow the flow of missions scheduled in the current day's ATO, determine what targets are planned for attack, and determine where strike aircraft are in the flow (e.g., takeoff at a certain time, on tanker, over target, returning to base) and their current state (e.g., fuel level and weapons remaining).

There is no automated tool to assist in describing the relationship between TCT nominations and those targets on the JIPTL and as-

---

[17]The *ISR-ICSP* discusses the operations and ISR integration efforts required to create a single integrated picture (ASD(C3I), 2000a).

[18]Personnel are required to hand carry information in hardcopy or softcopy formats (e.g., spreadsheet printouts, floppy disks) because the electronic transfer of data between systems and/or offices is not available.

signed for attack on the ATO. Such a capability is needed to support the tracking of the overall progress made by sorties flown against all targets in the area of interest and to enable the rerolling of strike assets from preplanned targets to emergent TCTs.

Last, the emergence of nonkinetic weapons and techniques (e.g., IW) requires modification of the traditional processes in the air operations center in order to integrate these techniques into the overall attack plan with sorties delivering kinetic bombs and missiles. A tool is needed to incorporate nonkinetic weapons and techniques into the overall attack plan.

RAPID WEAPON AND TARGET PAIRING

The lack of automated techniques and tools to help in the pairing of available weapon systems with a nominated TCT target increases the amount of time from initial detection to weapons on target; this is a critical variable in TCT operations.

Experiments aimed at demonstrating capabilities to perform the function of automated pairing of multiple attack options with emergent targets show that this function directly depends on the other functional areas that precede it. Capability shortfalls in the preceding functional areas need to be resolved before proposed automated pairing aids can fully contribute to streamlining the targeting process.

TIMELY DECISION AND ATTACK ORDER DISSEMINATION

Tactical communications—specifically, TADILs and mechanisms (e.g., gateways) to exchange data between different types of data communications—are clearly an enabler of this functional area.

A number of TADILs have been developed over many years to support near-real-time exchange of data among tactical data systems. The most recent is Link 16. Link 16 is an encrypted, jam-resistant, nodeless tactical digital data link network established by JTIDS-compatible communication terminals that transmit and receive data messages in the TADIL J message catalog.

Over the next several years, a large percentage of fighter aircraft will be equipped with Link 16 capabilities; other TCT attack assets and C2BM nodes will be equipped with other data communication capabilities (e.g., Situational Awareness Data Link, Improved Data Modem). Appropriate gateways must be built that support information exchange between these data communication capabilities. This will ensure the rapid dissemination of attack orders to all available assets and the exchange of information among C2BM nodes and between these nodes and attack assets. Close monitoring of the implementation of new data link communications is required to ensure that such systems are compatible or can work within whatever gateway framework that the Air Force eventually defines and adopts.

Similarly important is the need to ensure that Air Force–specific data link communications and C2BM processes are compatible with those of other services. Joint capabilities must be further refined, integrated, and implemented to support automated, cross-service, and component interactions.

RAPID ASSESSMENT OF WEAPON DELIVERY EFFECTS

The established procedures for conducting combat assessment of operations against deliberately planned targets—which consists of preplanned tasking of sensors to collect post-strike data for bomb impact assessment, battle damage assessment, munitions effects assessment, and reattack recommendations—are not sufficiently responsive to support combat assessment of attacks against mobile ground targets.

At present, there is no automated relay of attack data from the attack platforms to the various decision nodes charged with orchestrating the overall attack campaign. For example, voice pilot inflight reports are cumbersome and not adequate to support time dynamic decisions in TCT strike operations. Similarly, the real-time update of target lists to permit application of joint fires across the battlespace by multiple component commanders suggests a need for rapid reporting of the operational status of targets.

Some sort of quick-look, post-strike information to cover TCT strike locations—both from sensors on board attack assets or on accompanying attack assets and from dynamically retasked offboard sensors—must be rapidly provided to the TCT cell or other appropriate functionality embedded in the AOC.

Assuming that post-strike data is provided rapidly to the AOC, no automated tool(s) currently exist to help intelligence officers, combat planners, and combat operators collaboratively conduct rapid assessment of attacks against TCTs and rapidly restrike those targets if needed. As a result, more weapons may be expended against TCTs that pose a serious threat to U.S. and coalition forces than would be necessary if rapid assessment were performed.

INFRASTRUCTURE

To rapidly perform all the C2BM functions within the six functional areas discussed above, key personnel should be linked in an automated collaborative environment that includes tools and capabilities to rapidly exchange information. That exchange could be via voice, data, graphic, multimedia, or synthetic formats—whatever is specified by the user—and in a form that best fits the situation. This is particularly important if C2BM functions for TCT prosecution are distributed among several locations and nodes, as may be the case in the early phases of a rapidly emerging crisis, or when a forward AOC with a small footprint is used.

Today email, secure phones, and periodic video teleconferencing sessions, augmented by physical interaction, all support collaboration within the AOC and among geographically dispersed nodes supporting the AOC. Experimentation with collaborative tools (e.g., InfoWorkSpace and Collaborative Virtual Workspace) has demonstrated the need for improving the collaborative environment to better facilitate the exchange of combat information for more-dynamic operations. Results of those experiments indicate that the following improvements are needed to create a robust collaborative environment:

- Adequate tools to permit simultaneous interactions among personnel in more than one chat room.

- An ability to monitor multiple chat rooms or interphone communication channels.

- Applications supporting interactions of many people within a collective chat room.

- On-demand, sufficient, and assured communications.[19]

- Responsive, robust network and server architectures and management.

- Clearer definition of the required levels of information integration required to support collaborative operations across organizational and functions constructs.

An effective network manager and an empowered information manager are also essential to create a robust collaborative environment. In addition, the information manager is needed to ensure that all available data and information necessary to perform AOC functions are correlated and displayed in a user-friendly manner.

Finally, the Theater Battle Management Core System (TBMCS) is a potential major component in future DC2BM of TCT operations. As stated in the TCT capstone requirements document, the long-term objective is to have TCT functional capabilities integrated into TBMCS. However, the document recognizes that developmental and fielding issues with the current version of TBMCS may necessitate some separate TCT functional capabilities in the near term.[20] Thus, the suggested actions highlighted in this and the preceding sections of this chapter should be pursued in close coordination with TBMCS improvements so that TCT functionality and TBMCS developments are synchronized. The synchronization should ensure that required TCT functional capabilities brought to the AOC through near-term improvement programs can be integrated with the TBMCS version of record. In parallel, replacing or discarding TBMCS applications that are found lacking should also be addressed as the Air Force continues its evolutionary development and fielding of the TBMCS capability.

---

[19]A perennial complaint of warfighters is that existing communications capabilities do not support their current needs. Determining the best approach for ensuring that on-demand, sufficient, and assured communications are available to support the collaborative environment is beyond the scope of this study. Further research and experiments are needed.

[20]See the specific requirements in AC2ISRC, 1999.

Developing a Web-enabled TBMCS construct in the future is necessary to rapidly link and incorporate other applications and to permit electronic transfer of data necessary to perform DC2BM functions quickly. As steps in this direction continue, an overall automation standard and specific levels of compliance for each application or major data transfer should be defined to prevent any unintended delays. When many applications are assembled to support an end-to-end TCT functionality, it is expected that all applications will perform synergistically to meet user timeline goals. To ensure that such goals can be met, an overall system integrator for TBMCS and TCT-specific applications and systems is needed. The system integrator would test end-to-end functional performance of new versions of TBMCS and other stand-alone systems before they are fielded.

ORGANIZATION AND CONOPS

Our review of operations, exercises, and experiments over the past decade indicates that there is a continuing debate as to who should perform the TCT functions, and where and how they should be performed. To bring the various new programs and initiatives into a working construct requires the development of a TCT operational concept(s) that, at a minimum, defines the DC2BM functions that must be performed, where they are to be performed, and by whom. Preceding sections discussed the DC2BM functions necessary for successful TCT prosecution; the discussion in this section focuses on the remaining issues related to organization and CONOPS.

Several efforts are being made to define and institutionalize new DC2BM concepts for future combat operations.[21] Three general constructs or combinations seem to warrant further consideration for future joint force air component commander (JFACC) exercise of DC2BM of TCT operations: (1) a more integrated and more capable airborne Iron Triad (AWACS, Rivet Joint, and a properly equipped JSTARS), with robust data links to supporting elements in the AOC; (2) a distributed TCT functionality with specific functions performed at several locations under the control of the AOC and JFACC; and (3)

---

[21]AC2ISRC, 2000c; AC2ISRC/AFC2TIG, 2000a, 2000b; AC2ISRC/CC, 2000; ACC, 2000a, 2000b; AFDC/DR, 2000; COMUSAFE, 2000; USAF, 1999a, 1999b, 2000.

a fully integrated TCT cell or functionality within the AOC combat operations division.

An enhanced Iron Triad has the potential to provide robust DC2BM capabilities in small-scale conflicts or in very focused operations in a major theater war—with a smaller forward or theater presence than an AOC. The combination of an AWACS with an onboard airborne command element (ACE) and a Rivet Joint has proven DC2BM capabilities in air-to-air operations. However, it has yet to be determined what specific DC2BM capabilities should be fielded on board Iron Triad elements for air-to-ground TCT missions.

A grass roots effort initiated by the commanders of the JSTARS, AWACS, and Rivet Joint wings is addressing this issue.[22] This effort is leveraging the lessons from operational employments over the past decade and Air Force experimentation, possibly leading to the larger discussion of the relationship of the airborne C2BM nodes to the AOC structure currently being baselined.[23]

An ACE-like capability on board the Iron Triad should be considered for dynamic C2BM of large integrated strike/SEAD packages employing kinetic means and new IW techniques against targets defended by advanced air defenses.[24] It should be noted, however, that if the Iron Triad is to be used in air-to-ground TCT missions, considerable infrastructure and force structure are needed to maintain near-continuous operations.

An alternative airborne concept to the Iron Triad is a new multi-mission aircraft[25] that could potentially embed complete DC2BM

_____

[22]93rd ACW, 1998, 2000a, 2000b.

[23]Joint Staff/J6I, 2000.

[24]After we completed our research, the initial class of the C2ISR Package Commander course graduated at the AWACS wing. In addition, a C2ISR-focused Red Flag exercise was held early in 2001. Employment of an expert package commander is likely to result in better integration of ISR and C2 assets in large integrated strike/SEAD missions and other TCT missions.

[25]During the course of this analysis, the commander of Air Combat Command presented a proposal advocating an MMA that would be used to support the Expeditionary Air Force more fully, incorporating capabilities of AWACS, JSTARS, Rivet Joint, ABCCC, and adding an airborne information fusion functionality (Jumper, 2001). In addition, the AC2ISRC is developing an operational concept for multimission C2ISR

functionality of air-to-air and air-to-ground missions if the necessary requirements are included in its development. Note that an MMA may also reduce infrastructure and force structure requirements.

The second possible TCT construct is the creation of TCT functional capabilities that are distributed among several locations, e.g., AOC and forward TCT cell. Experiments with such a construct were done in EFX 1999 and JEFX 2000. The consensus from those experiments was that the TCT cell and the AOC should be collocated to more easily solve control issues and to facilitate interactions that arise in dynamic operations.[26] This view was largely due to the lack of a robust, collaborative environment with adequate tools; on-demand, sufficient, and assured communications; and a well-orchestrated server and network architecture and management within which to fully evaluate distributed operations. If the ability to establish and maintain such a robust, collaborative environment is questionable, further improvements in collaborative capabilities are needed before this construct can be embraced or rejected. The ability to establish and maintain a robust, collaborative environment is important not only for TCT operations, but for all distributed operations that the Air Force is seeking to enable (e.g., rear and forward AOC operations, AOC and airborne C2 nodes).

The third construct is the embedding of all TCT functionalities within the AOC, with the addition of necessary personnel and tools to perform the DC2BM of ISR and weapon system assets. This construct facilitates a tightly coupled C2BM process and the integration of deliberate and dynamic operations, with minimal risk to communication outages or limitations. However, without substantial automation of the C2BM process, this option may increase the footprint size of the AOC.[27] Furthermore, it is not entirely clear that physical collocation will solve the various challenges identified, nor

---

aircraft (AC2ISRC, 2000d). More recently, the AC2ISRC concept has furthered evolved into the multisensor command and control aircraft (MC2A) (AC2ISRC, 2001).

[26]AC2ISRC/AFEO, 2000b.

[27]While the total impact of embedding a TCT cell within the AOC has not been fully scoped, varying cell schemes and proposals are being reviewed and exercised at flag exercises to determine how best to imbed the TCT cell functionality within the overall AOC construct.

is it clear that limiting AOC structural options to a single alternative is desirable.

Combinations of the three preceding TCT constructs are also possible, and perhaps more likely. The combination of an airborne strike command element on JSTARS and a TCT cell forward was tested in JEFX 2000. We did not have sufficient data from that demonstration to recommend a particular construct, but we note the potential for such a construct, depending on the situation encountered and the level of effort anticipated for attack operations. What the demonstration does indicate is that perhaps a task-configurable TCT functionality should be pursued, based on desired capabilities and effects to be generated. This would allow the JFACC to organize and equip the AOC as he sees fit to respond to the situation—incorporating a range of capabilities, tools, and applications with which to achieve the desired effects.

Of course, providing complete flexibility will present an acquisition challenge to design an open architecture system capable of changing from a standard baseline into the form desired by the JFACC. Regardless of what construct is selected, personnel well-trained in DC2BM and deliberate AOC operations will be essential, as will flexible tools and synthetic environments within which to create the appropriate structure necessary to accomplish mission goals.

When examining all of these constructs, one must consider two mission-specific challenges. The first is to ensure that counterair missions against cruise and tactical ballistic missiles are well-integrated with all missions conducted in support of the area air defense commander (AADC) in a TAMD defense-in-depth concept. This is particularly important if the AADC is not in the AOC or if the JFACC is not the AADC. A second, related issue that needs further exploration is whether traditional AADC responsibilities, as exercised in the past by the JFACC, will remain applicable in the defense-in-depth concept in a fully networked environment. This study did not fully explore these issues. We present them as an adjunct to the AOC organizing structure discussion to show that additional doctrinal issues may surface as various structures are considered and folded into the AOC baseline definition.

Chapter Four

TOP-LEVEL SUGGESTED ACTIONS AND CONSIDERATIONS

TOP-LEVEL DC2BM SUGGESTED ACTIONS

Our top-level suggested actions to enhance DC2BM fall into four areas:

- Refinement and formalization of a CONOPS and TTP for DC2BM of ISR assets and weapons systems.

- Creation of a consistent, common view of the battlespace.

- Automation of the DC2BM process.

- Development of a robust, collaborative, distributed environment.

An updated and refined joint CONOPS and TTP for DC2BM of ISR assets and weapon systems for the prosecution of TCTs should be formalized.[1] Those CONOPS and TTP should address who, where, and how the DC2BM functions should be performed to support attacks

---

[1]Air Combat Command has promulgated a top-level CONOPS for C2 against TCTs for the Combat Air Forces (ACC, 1997). The Air Force Command and Control Training and Innovation Group also developed baseline CONOPS and TTP for TCTs to support JEFX 2000 (AC2ISRC/AFC2TIG, 2000a, 2000b). U.S. Joint Forces Command's explorations of approaches against TCTs in its Attack Operations Against Critical Mobile Targets (AOACMT) efforts, conducted in 1999–2000, contain various lessons that may also be applicable. Other service activities (e.g., Fleet Battle Experiment results) hold sources of documented insights that may be helpful. These documents provide a good starting point.

against cruise missiles, theater ballistic missile launchers, fielded forces, and SAMs—all using hide, survival, and shoot-and-scoot tactics. Because counter-TCT operations may be conducted in a wide range of crises (small- and large-scale operations) and because command authority prerogatives on how and where to exercise DC2BM of ISR and weapon systems to achieve desired effects may be vested at varying levels (from JFACC to National Command Authorities), the challenge lies in crafting a scalable and flexible TCT functionality. For example, the TCT functionality could be performed on board the Iron Triad; in a TCT cell embedded in the AOC; or at distributed locations.

An overarching DC2BM need in the mission areas of TMD, SEAD, and interdiction of small-unit ground forces is the refinement and mechanization of the processes, TTP, and systems for developing a consistent, common view of the battlespace that underpin the creation of common operational and tactical pictures. Without proper mechanization (including substantial automated support) to create an IPB-based geographic product underlay and subsequently to insert, with date and time stamps, the locations and dispositions of friendly, enemy, and neutral forces and targets (based on correlated multi-INT and surveillance data), decisionmakers will continue to be hampered in their performance of key functions (e.g., the timely deconfliction of TCT and preplanned missions and airspace that is implicit in rapid targeting, target nomination, and weapon and target pairing). Most elements[2] to achieve this DC2BM capability are available or are being developed, but they must be integrated and adequately funded.

Timelines for performing DC2BM functions—from target detection to weapons on target—must approach 10 minutes or less to achieve even modest effectiveness against many types of TCTs. To shorten current timelines, extensive automation (tools and applications) is required in each of the six functional areas. New DC2BM capabilities should be pursued in close coordination with TBMCS improvements.

---

[2]For example, Integrated Collection Management advanced concept technology demonstration, Automated Assistance with Intelligence Preparation of the Battlespace, Air Defense System Integrator, Dynamic Moving Target Indicator Experiment, TBMCS Situational Awareness and Assessment, and Global Command and Control System common operational picture.

One area for consideration within the TBMCS construct is the use of an open architecture structure leveraging and based on a Web-enabled[3] browsing and information access approach. Making use of disparate tools and applications of legacy, current, and future iterations may require a software translator or some sort of software wrapper.[4] Such considerations will be necessary if continued integration and automation of TBMCS core processes are pursued.

Finally, a robust, collaborative, distributed environment—tools for collaboration; on-demand, sufficient, and assured communications; a flexible network and server architecture with responsive operating protocols; an effective network manager; and an empowered information manager—is essential to the DC2BM process. An improved InfoWorkSpace-like tool with expanded simultaneous multinetwork participation capabilities and a stable network and server architecture needs to be developed. Formal network and information manager positions should be created for the AOC, and personnel must be trained to perform the functions of those positions. Finally, tests of the DC2BM process and tools in collaborative environments should be continued to determine both the extent to which decisionmakers will be able to rely on a collaborative, distributed environment in future operations and the level of personnel expertise and training that will be required to operate in such an environment.

TOP-LEVEL CONSIDERATIONS

In developing a flexible TCT functionality and selecting specific tools to support the timely performance of the six top-level DC2BM functional areas, four important factors should be considered:

---

[3]The term *Web-enabled* refers to the use of markup languages (e.g., HTML, XML) to present information using links between various documents, images, and servers within a given information enterprise accessed by a browser (e.g., Microsoft Internet Explorer, Netscape Navigator).

[4]A *software wrapper* refers to the technique of integrating two applications written in differing source code languages through the use of a software program that provides the translation services between the two as well as permitting a common user access environment to both applications.

- Tension between planned and dynamically tasked missions.
- Jointness and defense in depth.
- Balanced and synchronized investments among DC2BM, ISR, aircraft, and weapons.
- Robustness and flexibility.

First and most important, new DC2BM improvements should be designed so that they do not jeopardize Air Force capabilities to execute deliberately planned strike and air superiority missions. Steady-state deliberate air operations (those planned and then executed as planned in the ATO) will continue to be important in the application of air power (e.g., strike missions against fixed components of adversary centers of gravity). Well-established Air Force dynamic air-to-air capabilities with ATO preallocated assets (e.g., offensive counterair sweep operations and defensive counterair combat air patrol) will also continue to be important. These operations are designed to achieve the operational objectives of the overall air campaign plan. However, improper design of new DC2BM capabilities (lack of adequate automation, poor deconfliction aids, excessive fragmentation of responsibilities, and inadequate dynamic resource reallocation decision aids) may cause AOC leadership to devote too much attention to TSTs, to the detriment of more-deliberate strike and preassigned air superiority missions. Moreover, in the target-rich and weapon-constrained environment typical of a limited-warning major theater war, the unwise reallocation of weapons systems from preplanned targets (missions that nominally have a high likelihood of success) to TCTs (mission with a much lower likelihood of success) may disrupt the intended battle rhythm and the timely accomplishment of intended effects. This suggests that new DC2BM programs must interact seamlessly with key elements of TBMCS that are integral to deliberate operations—intelligence and operational databases, master air attack planning, and targeting tools.

Joint DC2BM solutions will be required in all of the mission areas we considered, but this is particularly clear in interdiction of small-unit ground forces and theater air and missile defense. Interdiction of small-unit ground forces clearly requires a joint response (as our experience in Operation Allied Force showed). In particular, ground forces and/or sensors must be involved to help detect, identify, and engage the targets under restrictive rules of engagement.

Theater air and missile threats also require a joint response, but for different reasons. Because it is clear that a single defense element cannot achieve the necessary robustness or performance level (i.e., 90 percent or greater probability of success), a joint defense-in-depth architecture, with more modest performance (e.g., 50 percent) in each layer, is usually envisioned for addressing cruise missiles and TBMs. The capabilities discussed above must consider such multi-layer defense constructs. For example, Air Force TCT CONOPS must address defense in depth; collaboration tools should be developed with such constructs in mind; and communications capabilities should be interoperable to ensure information can be passed among the multiple layers.

The third consideration underscores the point that investments in DC2BM must be synchronized with those in ISR, aircraft, and weapons, as well as being adequately funded (often not the case), to ensure that desired TCT capabilities are met. By this we mean if particular improvements in ISR, aircraft, and weapons are made to address TCTs, the necessary DC2BM programs must be included. For example, if a survivable, deep-look, long-dwell, focused-look surveillance capability consisting of several sensors is developed, the accompanying (and interoperable) DC2BM processes, systems, and personnel must also be developed. Alternatively, if a single sensor system is to be the focus of such a capability (for instance, space-based radar), the needed DC2BM must be included as part of the system. The challenge here is to ensure that major ISR, aircraft, and weapon program funding requirements are presented to decision-makers in their totality. DC2BM requirements should be included in the C4ISR plan required by DoD for the acquisition of major systems; the plan can then be used to justify funding of needed C2BM programs.

Last, although the United States currently does not face any major military threat to its national security against which it cannot ultimately prevail, uncertainties about future threats underscore the need for a hedging strategy. The U.S. desire to maintain superior capabilities to perform military missions makes such a strategy even more important. These conditions suggest that DC2BM programs and initiatives to fix current shortfalls must be sufficiently flexible to support more-advanced ISR, aircraft, and weapon systems. For example, efforts to improve rapid targeting of Global Positioning System (GPS)–aided weapons should be extendible to support GPS-aided standoff weapons with target acquisition sensors (such as Joint Air-to-Surface Standoff Missile and possibly advanced long-range missiles with inflight retargeting capabilities). Similarly, the design of rapid targeting and target nomination, target and weapon pairing, and order development and issuance for nonstealthy aircraft should not impede integration of comparable processes for low-observable aircraft and potential unmanned combat air vehicles. These considerations suggest that DC2BM programs should be carefully reviewed to see how well they support new ISR, aircraft, and weapon system programs.

REFERENCES

552nd ACW (Air Control Wing), 2000, informal paper defining C2ISR package commander, 552 ACW/552 OSS, Tinker AFB, Okla.

93rd ACW, 1998, "Draft Tactics Techniques and Procedures (TTP) for 93rd Air Control Wing Joint STARS Attack Support," Warner-Robins AFB, Ga., 15 February.

_____, 2000a, "Airborne Time Critical Targeting Cell TTP," Annex to AC2ISRC/AFC2TIG, 2000b, Warner-Robins AFB, Ga., July.

_____, 2000b, Airborne TCT Working Group meeting held at the 93rd ACW (JSTARS wing), Warner-Robins AFB, Ga., November.

AC2ISRC (Aerospace C2 & ISR Center), 1999, "Defeating Theater Time Critical Targets," Capstone Requirements Document (CRD), AC2ISRC (USAF) 401-98, draft, Langley AFB, Va., August. This document was superseded by the Family of Systems Requirements Document, AC2ISRC, 2000a.

_____, 2000a, *Defeating Theater Time Critical Targets*, Family of Systems Requirements Document (FSRD), AC2ISRC (USAF) 401-98, Langley AFB, Va., 11 January.

_____, 2000b, *Strategy and Modernization Plan for Time Critical Targeting and Real Time Information to the Cockpit*, Langley AFB, Va., 7 February.

_____, 2000c, "Draft Concept of Operations for Expeditionary Operations Center," Langley AFB, Va., 21 March.

_____, 2000d, "Multi-Mission Command and Control, Intelligence, Surveillance, and Reconnaissance (C2&ISR) Aircraft Operational Concept," draft, Langley AFB, Va., 30 October.

_____, 2001, "Concept of Operations for the Multi-Sensor Command and Control Aircraft," draft, Langley AFB, Va., 25 April.

AC2ISRC/AFC2TIG (Air Force Command and Control Training and Innovation Group), 2000a, "Baseline Concept of Operations for Time Critical Targeting (TCT) in JEFX 2000," Version 6.0, Hurlburt Field, Fla., 2 September.

_____, 2000b, "Tactics, Techniques, and Procedures (TTP), Manpower and System Requirements for Time Critical Targeting (TCT) in JEFX 2000," Version 6.2, Hurlburt Field, Fla., 2 September.

AC2ISRC/AFEO (Air Force Experimentation Office), 2000a, "JEFX 2000 Detailed Experiment Plan (DEP)," Langley AFB, Va.

_____, 2000b, "JEFX 2000 Final Report," draft, Langley AFB, Va., December. Government publication; not releasable to the general public.

AC2ISRC/C2B (Command and Control Battlelab), 2000a, "CONOPS for Intelligence, Surveillance, Reconnaissance Battle Management (ISR BM) in JEFX 2000," Version 2.0, Hurlburt Field, Fla., July.

_____, 2000b, "ISR BM Initiative White Paper: Abstract for JEFX 2000," Hurlburt Field, Fla.

AC2ISRC/C2N (Strategic Forces Directorate), 2000a, "Time Critical Targeting Automated TCT Tools," briefing charts, Langley AFB, Va.

_____, 2000b, "Joint Integrated TCT cell," briefing charts, Langley AFB, Va., July.

AC2ISRC/CC (Aerospace C2 & ISR Center Commander), 2000, "CAF AOC Weapon System," briefing charts, Langley AFB, Va., April.

AC2ISRC/JWID JPO (JWID Joint Project Office), 2000a, "Joint Warrior Interoperability Demonstration (JWID) 2000–2001 Update," briefing charts, Langley AFB, Va., November.

_____, 2000b, "Joint Warrior Interoperability Demonstration (JWID) 2000-2001 Campaign Plan (JWID '00–01 CPLAN)," Langley AFB, Va., March.

ACC (Air Combat Command), 1996, "Airborne Warning and Control System (AWACS) Requirements Roadmap," HQ ACC/DRC, Langley AFB, Va., May 16.

_____, 1997, *Combat Air Forces Concept of Operations for Command and Control Against Time Critical Targets*, Langley AFB, Va., 8 July.

_____, 1999, "Draft Combat Air Forces Concept of Operations for Command and Control of Intelligence, Surveillance and Reconnaissance Operations," Langley AFB, Va., 28 October.

_____, 2000a, "Draft Combat Air Forces Control and Reporting Center Concept of Operations for Expeditionary Air Operations," Langley AFB, Va., 14 June.

_____, 2000b, "Draft Air Force Concept of Operations for Aerospace Operations Center," Langley AFB, Va.

_____, 2000c, "Strategic Vision for Command & Control," briefing charts, Langley AFB, Va.

AFDC/DR (Air Force Doctrine Center/Requirements Directorate), 2000, "Draft Aerospace Commander's Handbook: The JFACC," 3 October 2000 prototype, Maxwell AFB, Ala., October.

ASD(C3I) (Assistant Secretary of Defense [Command, Control, Communications, and Intelligence]), 2000a, *Intelligence, Surveillance, and Reconnaissance Integrated Capstone Strategic Plan (ISR-ICSP)*, Version 1.0, Washington, D.C., 3 November. Government publication; not releasable to the general public.

_____, 2000b, "Annex E, ISR Integration and Challenges Ahead (Draft)," Supplement to Chapter 7 of the *ISR-ICSP*, Version 1.0, Washington, D.C., 31 May.

Boyle, Col Edward J., 1999, "Operation Allied Force: Rapid Targeting, Collection, and Execution," briefing charts, USAFE/IN, Ramstein AB, Germany, 4 August. Government publication; not releasable to the general public.

COMUSAFE (Commander, U.S. Air Forces in Europe), 2000, "Limitations of Doctrine," briefing charts, USAFE, Ramstein AB, Germany.

DoD (Department of Defense), 2001, *Space-Based Radar Roadmap*, report to Congress, Washington, D.C., 1 May. Government publication; not releasable to the general public.

Dupont, Daniel G., 2000, "OSD Mulling Funding for Single Integrated Air Picture Development," *Inside the Air Force*, Vol. 11, No. 48, 1 December , pp. 3–4.

Hamilton, Thomas, and Richard Mesic, 2001, "A Simple Game-Theoretic Approach to SEAD and Other Time Critical Target Analyses," internal document, RAND. Not releasable to the general public.

Hura, Myron, Gary McLeod, et al., 2000, *Interoperability: A Continuing Challenge in Coalition Air Operations*, MR-1235-AF, RAND.

Hura, Myron, Gary McLeod, et al., 2002, *Dynamic Command and Control: Focus on Time Critical Targets*, RAND. Government publication; not releasable to the general public.

JC2ISR JT&E, 2000, JC2ISR JT&E Joint Working Group meeting, Joint Warfighter Facility, Suffolk, Va., 28-30 November.

Joint Staff/J6I, 2000, "Global Information Grid: Supporting the 21st Century Warrior," briefing charts, Washington, D.C., 1 December.

JTAMDO and BMDO, 1998, *1998-B Theater Air and Missile Defense (TAMD) Master Plan*, Washington, D.C., December. Government publication; not releasable to the general public.

Jumper, Gen John P., 2001, "Global Strike Task Force Way Ahead," briefing, Air Combat Command, Langley AFB, Va., February.

Mayer, MSgt Daryl, 2001, "Upgraded AWACS Declared Ready for Duty," news release, Electronic Systems Center Public Affairs, Hanscom AFB, Mass., 25 June.

OASD(C3I)/DASD(C3ISR&S), Joint Staff/J2, and Joint Staff/J6, 2000, "Information Superiority Investment Strategy (ISIS) Options

Description Document," Washington, D.C., 31 October. Government publication; not releasable to the general public.

Ochmanek, David A., Edward Harshberger, David A. Thaler, and Glenn A. Kent, 1998, *To Find and Not to Yield: How Advances in Information and Firepower Can Transform Theater Warfare*, MR-958-AF, RAND.

Sauer, Philip S., William A. Williams, et al., 2000, "Joint Capability Assessment of Joint Expeditionary Force Experiment 2000," internal document, RAND, November. Not releasable to the general public.

USAF (U.S. Air Force), 1999a, "Operational Procedures: Aerospace Operations Center," Air Force Instruction 13-1AOC, Volume 3, Washington, D.C., 1 June. Currently under revision.

_____, 1999b, "Draft Expeditionary Air Force C2 Process Manual," Washington, D.C., December.

_____, 2000, "Draft Aerospace Operations Center Process Manual: Air Force Tactics, Techniques, and Procedures," Washington, D.C., 1 August.

USAF/XOI (U.S. Air Force/ISR Directorate), 2000, "USAF C2ISR Battle Management Initiatives," briefing charts, Washington, D.C.

USJFCOM/J9 (U.S. Joint Forces Command/Joint Experimentation Directorate), 1999, "A White Paper for Attack Operations Against Critical Mobile Targets Concept," Suffolk, Va., 9 November.

_____, 2000a, "Attack Operations 2000," briefing charts and experiment demonstration documents, Suffolk, Va.

_____, 2000b, "Final Report Intelligence, Surveillance & Reconnaissance Workshop November 30–December 2, 1999, Attack Operations Against Critical Mobile Targets (AOACMT) Concept," Suffolk, Va.

Wolfe, Frank, 2001, "Mushala: Revised F-22 Buy Plan to Allow Institution of Cost Reduction Initiatives," *Defense Daily*, 24 September , p. 4.